BUSINESS and FINANCIAL STATISTICS USING MINITAB 12 and MICROSOFT EXCEL 97

John C. Lee
The Chase Manhattan Bank, New York

World Scientific
Singapore • New Jersey • London • Hong Kong

Published by

World Scientific Publishing Co. Pte. Ltd.

P O Box 128, Farrer Road, Singapore 912805

USA office: Suite 1B, 1060 Main Street, River Edge, NJ 07661

UK office: 57 Shelton Street, Covent Garden, London WC2H 9HE

British Library Cataloguing-in-Publication Data
A catalogue record for this book is available from the British Library.

BUSINESS AND FINANCIAL STATISTICS USING MINITAB 12 AND MICROSOFT EXCEL 97

Copyright © 2000 by World Scientific Publishing Co. Pte. Ltd.

All rights reserved. This book, or parts thereof, may not be reproduced in any form or by any means, electronic or mechanical, including photocopying, recording or any information storage and retrieval system now known or to be invented, without written permission from the Publisher.

For photocopying of material in this volume, please pay a copying fee through the Copyright Clearance Center, Inc., 222 Rosewood Drive, Danvers, MA 01923, USA. In this case permission to photocopy is not required from the publisher.

ISBN 981-02-3879-7 (pbk)

Printed in Singapore.

PREFACE

This book uses MINITAB 12 and Microsoft Excel 97 to show how these new computer programs can be applied in business and financial data analysis.

There are 22 chapters. The first 21 chapters closely follows the 2nd edition of the statistics book entitled <u>Statistics for Business and Financial Economics</u> by Cheng F. Lee, John C. Lee and Alice C. Lee. These chapters are divided into ***five portions:***

1) Introduction and Descriptive Statistics
2) Probability and Important Distributions
3) Statistical Inferences Based on Samples
4) Regression and Correlation: Relating Two or more Variables
5) Selected Topics in Statistical Analysis for Business and Economics

In addition, Chapter 22 shows how *Microsoft Excel 97* can be used to calculate option for individual stock, index option and foreign currency option.

There are two alternative ways to use this book:

1) It can be used together with the above mentioned statistics book, <u>Statistics for Business and Financial Economics</u>
2) It can be used independently by the reader to learn how to apply *MINITAB 12* and *Microsoft Excel 97* in statistical data.

I would like to express my thanks to Ms. Kim Tan for her editorial comments and suggestions. In addition, I would also like to thank Professor Cheng F. Lee and Dr. Ta-Peng Wu, From Rutgers University, for their comments in Chapter 22.

John C. Lee
Chase Manhattan Bank, New York
November 1999

TABLE OF CONTENTS

Introduction and Descriptive Statistics

Chapter 1   Introduction   1
    1.1 Statistics' Computational Intensity   1
    1.2 Environment of MINITAB   2
    1.3 Statistical Environment of Microsoft Excel 97   3
    1.4 Using This Book   8

Chapter 2   Data Collection and Presentation   9
    2.1 Understanding Statistics   9
    2.2 Analyzing GM's Data Using MINITAB   10
    2.3 Analyzing GM's Data Using EXCEL   16
    2.5 Statistical Summary   23

Chapter 3   Frequency Distributions and Data Analyses   24
    3.1 Introduction   24
    3.2 Tally in MINITAB   25
    3.3 Tally in EXCEL   25
    3.4 Dotplot   27
    3.5 Histogram   28
    3.6 Stem and Leaf   29
    3.7 Boxplot (Box-and-Whisker Plot)   30
    3.8 EXCEL Programming   34
    3.9 MINITAB Programming   36
    3.10 Statistical Summary   38

Chapter 4   Numerical Summary Measures   39
    4.1 Introduction   39
    4.2 Arithmetic Mean   39
    4.3 Median and Quartiles   42
    4.4 Standard Deviation   44
    4.5 Variance   45
    4.6 Z Score   48
    4.7 Skewness   51
    4.8 Statistical Summary   54

Probability and Important Distributions

Chapter 5   Probability Concepts and Their Analysis   56

- 5.1 Introduction   56
- 5.2 Probability   57
- 5.3 Permutations   59
- 5.4 Combination   64
- 5.5 Statistical Summary   69

Chapter 6   Discrete Random Variables and Probability Distributions   70

- 6.1 Introduction   70
- 6.2 Cumulative Density Function   72
- 6.3 Mean of a Discrete Variable   73
- 6.4 Variance of a Discrete Variable   76
- 6.5 Binomial Random Variable   78
- 6.6 Poisson Random Variable   82
- 6.7 Statistical Summary   84

Chapter 7   The Normal and Lognormal Distributions   85

- 7.1 Introduction   85
- 7.2 Uniform Distribution   85
- 7.3 Normal Distribution   91
- 7.4 Lognormal Distribution   100
- 7.5 Standard Normal Distribution   102
- 7.6 Normal Approximating the Binomial   103
- 7.7 Normal Approximating the Poisson   105
- 7.8 Statistical Summary   106

Chapter 8   Sampling and Sampling Distributions   107

- 8.1 Introduction   107
- 8.2 Uniform Distribution   109
- 8.3 Normal Distribution   113
- 8.4 Lognormal Distribution   116
- 8.5 Binomial Distribution   121
- 8.6 Poisson Distribution   124
- 8.7 Statistical Summary   128

Chapter 9   Other Continuous Distributions and Moments for Distributions   129

    9.1 Introduction   129
    9.2 t Distribution   129
    9.3 Chi-Square Distribution   131
    9.4 F Distribution   133
    9.5 Exponential Distribution   134
    9.6 Central Limit Theorem   136
    9.7 Statistical Summary   148

Statistical Inferences Based on Samples

Chapter 10   Estimation and Statistical Quality Control   149

    10.1 Introduction   149
    10.2 Interval Estimates for μ when σ^2 Is Known   150
    10.3 Confidence Intervals for μ when σ^2 Is Unknown   158
    10.4 Confidence Interval for the Population Proportion   161
    10.5 Confidence Intervals for the Variance   166
    10.6 Statistical Summary   167

Chapter 11   Hypothesis Testing   168

    11.1 Introduction   168
    11.2 One-Tailed Tests of Mean for Large Samples   169
    11.3 Two-Tailed Tests of Mean for Large Samples   172
    11.4 One-Tailed Tests of Mean for Small Samples   173
    11.5 Difference of Two Means   175
    11.6 Hypothesis Testing for a Population Proportion   178
    11.7 Statistical Summary   181

Chapter 12   Analysis of Variance and Chi-Square Tests   182

    12.1 Introduction   182
    12.2 One-Way Analysis of Variance   186
    12.3 Two-Way Analysis of Variance   196
    12.4 Chi-Square Test   200
    12.5 Statistical Summary   202

Regression and Correlation: Relating Two or More Variables

Chapter 13   Simple Linear Regression and the Correlation Coefficient   203

    13.1 Introduction   203
    13.2 Regression Analysis   203
    13.3 Coefficient of Determination   211
    13.4 Correlation Coefficient   212
    13.5 Regression Examples   215
    13.6 Statistical Summary   219

Chapter 14   Simple Linear Regression and Correlation: Analyses and Applications   220

    14.1 Introduction   220
    14.2 Two-Tail t Test for β 221
    14.3 Two-Tail t Test for α 223
    14.4 Confidence Interval of β 223
    14.5 F Test   225
    14.6 The Relationship Between the F Test and the t Test   226
    14.7 Predicting   226
    14.8 Regression Examples   228
    14.9 Statistical Summary   233

Chapter 15   Multiple Linear Regression   234

    15.1 Introduction   234
    15.2 R-Square   237
    15.3 F Test   238
    15.4 t Test   241
    15.5 Confidence Interval of β 242
    15.6 Predicting   244
    15.7 Another Regression Example   246
    15.8 Stepwise Regression   248
    15.9 Statistical Summary   249

Chapter 16 Other Topics in Applied Regression Analysis 250

16.1 Introduction 250
16.2 Linearity 256
16.3 The Expected Value of the Residual Term Is Zero 260
16.4 The Variance of the Residual Term Is Constant 261
16.5 The Residual Terms Are Independent 262
16.6 The Independent Variables Are Uncorrelated 264
16.7 The Residuals Are Normally Distributed 267
16.8 Stepwise Regression 268
16.9 Statistical Summary 269

Selected Topics in Statistical Analysis for Business and Economics

Chapter 17 Nonparametric Statistics 270

17.1 Introduction 270
17.2 Mann-Whitney U Test 271
17.3 Kruskal-Wallis Test 274
17.4 Wilcoxon Matched-Pairs Signed-Rank Test 277
17.5 Ranking 280
17.6 Statistical Summary 282

Chapter 18 Time-Series: Analysis, Model, and Forecasting 283

18.1 Introduction 283
18.2 Moving Averages 283
18.3 Linear Time Trend Regression 286
18.4 Exponential Smoothing 289
18.5 Holt-Winters 295
18.6 Statistical Summary 304

Chapter 19 Index Numbers and Stock Market Indexes 305

19.1 Introduction 305
19.2 Simple Price Index 305
19.3 Laspeyres Price Index 310
19.4 Paasche Price Index 312
19.5 Fisher's Ideal Price Index 315
19.6 Laspeyres Quantity Index 318
19.7 Paasche Quantity Index 321
19.8 Fisher's Ideal Quantity Index 323
19.9 Statistical Summary 323

Chapter 20 Sampling Surveys: Methods and Applications   324

- 20.1 Introduction   324
- 20.2 Random Number Tables   324
- 20.3 Confidence Interval for the Population Mean   326
- 20.4 Confidence Interval for the Population Proportion   330
- 20.5 Determining Sample Size   332
- 20.6 Statistical Summary   334

Chapter 21 Statistical Decision Theory: Methods and Applications   335

- 21.1 Introduction   335
- 21.2 Decisions Based on Extreme Values   335
- 21.3 Expected Monetary Value   337
- 21.4 Bayes Strategies   338
- 21.5 Statistical Summary   343

Chapter 22 Using Microsoft Excel to Estimate Alternative Option Pricing Models   344

- 22.1 Introduction   344
- 22.2 Option Model for Individual Stocks   344
- 22.3 Option Model for Stock Indices   346
- 22.4 Option for Currencies   348
- 22.5 Statistical Summary   350

Appendex A References   351

Index   352

BUSINESS and FINANCIAL STATISTICS USING MINITAB 12 and MICROSOFT EXCEL 97

CHAPTER 1
INTRODUCTION

1.1 Statistics' Computational Intensity

Statistics is a very intense computational discipline. We can get a sense of the computational intensity of statistics by looking at the average statistical formula and the correlation coefficient statistical formula shown below.

$$\bar{x} = \frac{\sum_{i=1}^{n} x_i}{n}$$

$$r = \frac{\frac{1}{n-1}\sum_{i=1}^{n}(x_i - \bar{x})(y_i - \bar{y})}{[\frac{1}{n-1}\sum_{i=1}^{n}(x_i - \bar{x})^2]^{1/2}[\frac{1}{n-1}\sum_{i=1}^{n}(y_i - \bar{y})^2]^{1/2}}$$

Average   Correlation Coefficient

Both of these formulas are relatively complicated and could give the statistician considerable computational trouble. The average formula looks fairly simple, but it can become difficult when n increases. The correlation coefficient formula is obviously more complex, and tedious to perform -- and other computational formulas are even more complicated. Many statistical formulas can take days or weeks by hand or even with a calculator. In the 'old days' before statistical software and computers, statisticians spent a great deal of time doing tedious, repetitive calculations and then tracking down the inevitable errors.

Now consider a situation in which the statistician must calculate the correlation coefficient more than once. Doing it even once by hand is daunting, and having to do it more than once can be overwhelming.

Statistics, of course, involves more than numbers alone. The presentation of information by means of graphs and pictures is an essential part of the discipline. Few people, however, have the time and skill to draw accurate, good-looking graphs.

Fortunately with the advent of personal computers, the availability of computer software has made statistical analysis easier. The computer software can do the following,

1. Do statistical calculations very quickly.
2. Do statistical calculations without mistakes.
3. Do repetitive calculations quickly.
4. Generate statistical graphs quickly and accurately.

In this book we will look at two software packages to do statistical analyses. The two software packages are MINITAB STATISTICAL SOFTWARE Release 12 and Microsoft Excel 97. MINITAB is a software package that is written specifically to do statistical analyses. Microsoft Excel 97 (EXCEL) is a powerful business application that has strong statistical features.

MINITAB's strength is that it is written to do specifically statistical analysis. This is why it is very easy to create a box plot in MINITAB. EXCEL's strength is that it has a very strong programming language called Visual Basic for Applications (VBA). To create a box plot in EXCEL, we would have to write a program using VBA to create a box plot. Even though doing this would be more difficult than using MINITAB, the reward would be worth the effort. We can transfer our knowledge in creating a box plot in EXCEL to other non-statistical EXCEL projects.

An important reason in using EXCEL is it has become a "standard" program in the business world. This makes it very important to study EXCEL.

1.2 Environment of MINITAB

MINITAB consists of two main parts, the *data window* and the *session window*.

The data window is where the statistical data are entered. The session window is where all statistical analysis are shown and where statistical commands called session commands can be issued. When you start MINITAB, the data window and session window are displayed by default. This is shown below.

INTRODUCTION 3

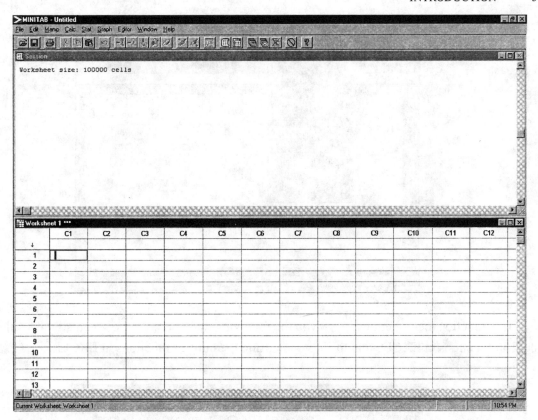

The session window is where statistical commands and reports are displayed. Many statistical calculations require several commands. With MINITAB you can group commands together and give a name to the group of commands. Such a group is called a *macro*. The user can execute the group of commands just by typing the group name. Grouping commands is computer programming. The capability to program is one of MINITAB's strengths.

The basic procedures of using MINITAB will be discussed in the next two chapters.

1.3 Statistical Environment of Microsoft Excel 97

EXCEL has many built in statistical features. First, there are many built in statistical functions. To view the statistical functions, go to the *Insert* menu and choose the *Function* menu item as shown below. Going forward we will use the following notation to get to menu items, **Menu → Menu item**. To get to the function menu we will then indicate **Insert → Function**.

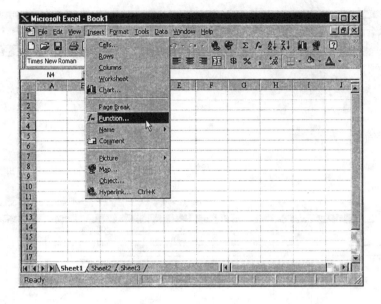

The Function menu item will give you the following dialog box to show you all the available built-in functions in EXCEL and to aid you in using the built-in functions in EXCEL.

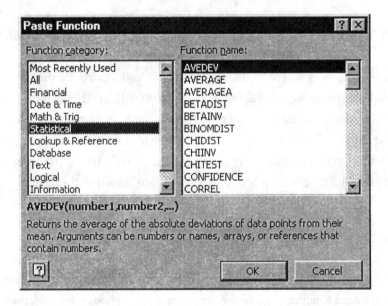

Below is a table showing all the available statistical functions in the above dialog box.

Statistical Functions

AVEDEV	Returns the average of the absolute deviations of data points from their mean
AVERAGE	Returns the average of its arguments
AVERAGEA	Returns the average of its arguments, including numbers, text, and logical values
BETADIST	Returns the cumulative beta probability density function
BETAINV	Returns the inverse of the cumulative beta probability density function
BINOMDIST	Returns the individual term binomial distribution probability
CHIDIST	Returns the one-tailed probability of the chi-squared distribution
CHIINV	Returns the inverse of the one-tailed probability of the chi-squared distribution
CHITEST	Returns the test for independence
CONFIDENCE	Returns the confidence interval for a population mean
CORREL	Returns the correlation coefficient between two data sets
COUNT	Counts how many numbers are in the list of arguments
COUNTA	Counts how many values are in the list of arguments
COVAR	Returns covariance, the average of the products of paired deviations
CRITBINOM	Returns the smallest value for which the cumulative binomial distribution is less than or equal to a criterion value
DEVSQ	Returns the sum of squares of deviations
EXPONDIST	Returns the exponential distribution
FDIST	Returns the F probability distribution
FINV	Returns the inverse of the F probability distribution
FISHER	Returns the Fisher transformation
FISHERINV	Returns the inverse of the Fisher transformation
FORECAST	Returns a value along a linear trend
FREQUENCY	Returns a frequency distribution as a vertical array
FTEST	Returns the result of an F-test
GAMMADIST	Returns the gamma distribution
GAMMAINV	Returns the inverse of the gamma cumulative distribution
GAMMALN	Returns the natural logarithm of the gamma function, G(x)
GEOMEAN	Returns the geometric mean
GROWTH	Returns values along an exponential trend
HARMEAN	Returns the harmonic mean
HYPGEOMDIST	Returns the hypergeometric distribution
INTERCEPT	Returns the intercept of the linear regression line
KURT	Returns the kurtosis of a data set
LARGE	Returns the k-th largest value in a data set
LINEST	Returns the parameters of a linear trend
LOGEST	Returns the parameters of an exponential trend
LOGINV	Returns the inverse of the lognormal distribution
LOGNORMDIST	Returns the cumulative lognormal distribution
MAX	Returns the maximum value in a list of arguments
MAXA	Returns the maximum value in a list of arguments, including numbers, text, and logical values
MEDIAN	Returns the median of the given numbers
MIN	Returns the minimum value in a list of arguments
MINA	Returns the smallest value in a list of arguments, including numbers, text, and logical values
MODE	Returns the most common value in a data set
NEGBINOMDIST	Returns the negative binomial distribution
NORMDIST	Returns the normal cumulative distribution
NORMINV	Returns the inverse of the normal cumulative distribution

CHAPTER 1

NORMSDIST	Returns the standard normal cumulative distribution
NORMSINV	Returns the inverse of the standard normal cumulative distribution
PEARSON	Returns the Pearson product moment correlation coefficient
PERCENTILE	Returns the k-th percentile of values in a range
PERCENTRANK	Returns the percentage rank of a value in a data set
PERMUT	Returns the number of permutations for a given number of objects
POISSON	Returns the Poisson distribution
PROB	Returns the probability that values in a range are between two limits
QUARTILE	Returns the quartile of a data set
RANK	Returns the rank of a number in a list of numbers
RSQ	Returns the square of the Pearson product moment correlation coefficient
SKEW	Returns the skewness of a distribution
SLOPE	Returns the slope of the linear regression line
SMALL	Returns the k-th smallest value in a data set
STANDARDIZE	Returns a normalized value
STDEV	Estimates standard deviation based on a sample
STDEVA	Estimates standard deviation based on a sample, including numbers, text, and logical values
STDEVP	Calculates standard deviation based on the entire population
STDEVPA	Calculates standard deviation based on the entire population, including numbers, text, and logical values
STEYX	Returns the standard error of the predicted y-value for each x in the regression
TDIST	Returns the Student's t-distribution
TINV	Returns the inverse of the Student's t-distribution
TREND	Returns values along a linear trend
TRIMMEAN	Returns the mean of the interior of a data set
TTEST	Returns the probability associated with a Student's t-test
VAR	Estimates variance based on a sample
VARA	Estimates variance based on a sample, including numbers, text, and logical values
VARP	Calculates variance based on the entire population
VARPA	Calculates variance based on the entire population, including numbers, text, and logical values
WEIBULL	Returns the Weibull distribution
ZTEST	Returns the two-tailed P-value of a z-test

In EXCEL it is possible to extend the functionality of EXCEL. This is done through computer programs called *add-ins*. Excel comes with a statistical add-in. If installed properly you would run the add-in by choosing **Tools** →**Data Analysis.** If it is not there, you would need to add the add-in by choosing **Tools** → **Add-in.** This would give you the following dialog box.

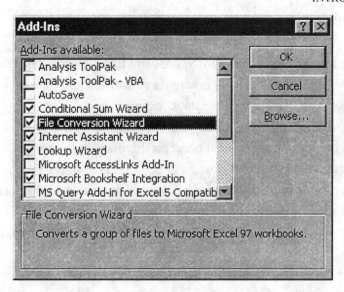

Check the second option, Analysis ToolPak - VBA , to install the statistical add-in.

After doing this you will see a Data Analysis menu item under the Tools menu. The Data Analysis menu item will give you the following dialog box,

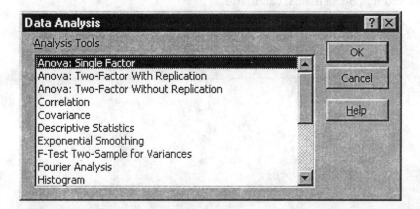

The Data Analysis dialog box shows all the statistical anlaysis available in EXCEL. To do a specific analysis of interest, highlight the statistical anlysis of interst and then click the *OK* button.

1.4 Using This Book

This book recognizes that the study of MINITAB and EXCEL cannot be separated from the study of statistics. Therefore, this book first looks at statistics and then shows how MINITAB and EXCEL can be used to solve specific statistical problems.

One strength of this book is that it analyzes the computer output created by MINITAB and EXCEL and then links it to the statistical concepts discussed in the chapter.

The book's organization follows Cheng F. Lee, John Lee, and Alice Lee's *Statistics for Business and Financial Economics* chapter by chapter. It quickly explores and reviews the statistical concepts for every chapter in Lee's book and then shows how MINITAB and EXCEL can be used to solve statistical problems, therefore making this book an excellent companion to Lee's book.

This book can also be used with other textbooks. Most of the topics discussed in other statistics texts are also discussed here. The only difference will be the organization.

What you might discover after studying MINITAB and EXCEL is a greater appreciation for statistics and that statistics is an exciting discipline. MINITAB and EXCEL allows the student of statistics to concentrate more on conceptual understanding, data analysis, real-world applications and less on statistical calculations.

CHAPTER 2
DATA COLLECITON AND PRESENTATION

2.1 Understanding Statistics

Understanding what statistics is will make it easier to study. In general, *statistics is a discipline that studies data and people usually makes decisions based on that analysis*. This is a very general definition. We will be more specific about what statistics is in Chapter 5.

Statistics studies data by either graphing the data and/or doing mathematical calculations on the data. In this book we will look at the different graphs and mathematical calculations used in statistics. We will use MINITAB and EXCEL to do the graphs and mathematical calculations.

Figure 2-1 Graphical Representation of Statistics and the Role of MINITAB and EXCEL

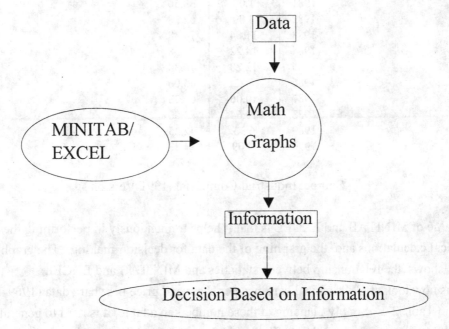

2.2 Analyzing GM's Data Using MINITAB

Table 2-1 Earnings Per Share and Price Per Share for General Motors (GM)

Year	Earnings Per Share	Price Per Share
1969	5.95	69.13
1970	2.09	80.05
1971	6.72	80.50
1972	7.51	81.13
1973	8.34	46.13
1974	3.27	30.75
1975	4.32	57.63
1976	10.08	78.50
1977	11.62	62.88
1978	12.24	53.75
1979	10.04	50.00
1980	-2.65	45.00
1981	1.07	38.50
1982	3.09	62.38
1983	11.84	74.38
1984	14.22	78.38
1985	12.28	70.38
1986	8.22	66.00
1987	10.06	61.38
1988	13.64	83.50
1989	6.33	42.25
1990	-4.09	34.38

Source: Industrial Compustat, 1991 Version

The value of MINITAB and EXCEL is that it helps tremendously in performing these mathematical calculations and the graphing of the data for decision-making. The graph in Figure 2-1 shows the relationship between statistics and MINITAB and EXCEL.

Let's us now look at GM's annual earnings per share and price per share data (1969-1990) as presented in Table 2-1. Simply glancing at these numbers in a table, it is hard to generalize or understand them. A graph can help us. We could draw the graph by hand or we could do it with MINITAB or EXCEL. We probably should choose MINITAB or EXCEL, because MINITAB or EXCEL can draw a graph a lot quicker than we can by hand.

Your MINITAB worksheet should look like the following after you have entered GM's data.

	C1	C2	C3	C4
	year	gmeps	gmpps	
1	1969	5.95	69.13	
2	1970	2.09	80.05	
3	1971	6.72	80.50	
4	1972	7.51	81.13	
5	1973	8.34	46.13	
6	1974	3.27	30.75	
7	1975	4.32	57.63	
8	1976	10.08	78.50	
9	1977	11.62	62.88	
10	1978	12.24	53.75	
11	1979	10.04	50.00	
12	1980	-2.65	45.00	
13	1981	1.07	38.50	
14	1982	3.09	62.38	
15	1983	11.84	74.38	
16	1984	14.22	78.38	
17	1985	12.28	70.38	
18	1986	8.22	66.00	
19	1987	10.06	61.38	
20	1988	13.64	83.50	
21	1989	6.33	42.25	
22	1990	-4.09	34.35	

We will graph the data. There are two ways to do analysis in MINITAB. One way is by using *menu commands* and the other by *session commands*. In this book we will emphasize the sessions command because they are important when writing MINITAB macros. To use session commands we will need to enable them by choosing **Editor → Enable Command Language** as shown below.

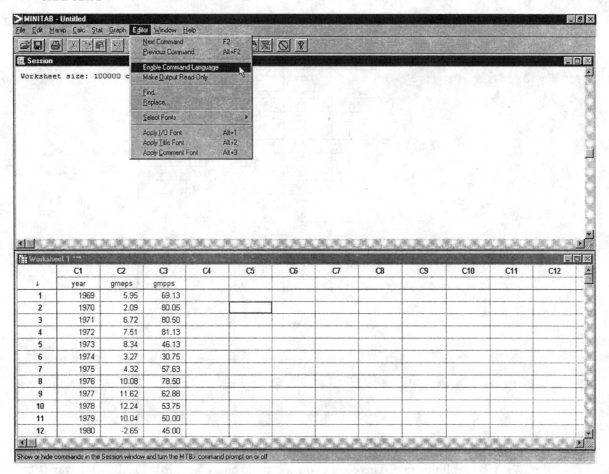

This will display the following command prompt in the session window.

MTB >

Let us now graph GM's earnings per share by typing the appropriate MINITAB commands in the session window as shown below.

DATA COLLECTION AND PRESENTATION   13

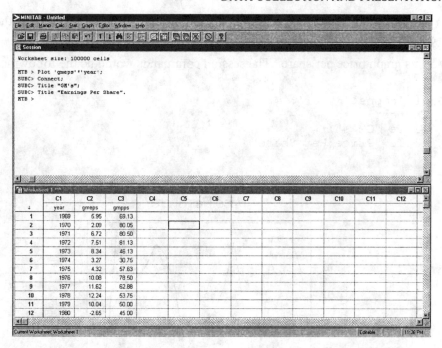

The MINITAB commands in the session window produces the following graph.

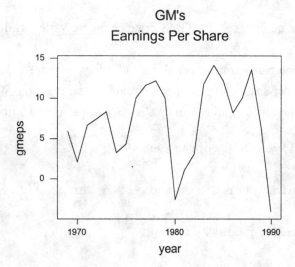

14   CHAPTER 2

From this plot we can quickly see that GM's earnings per share vary wildly for the years 1969 to 1990.

Now let us graph price per share. The session commands would be the following.

```
MTB > Plot 'gmpps'*'year';
SUBC>    Connect;
SUBC>    Title "GM's";
SUBC>    Title "Price Per Share".
```

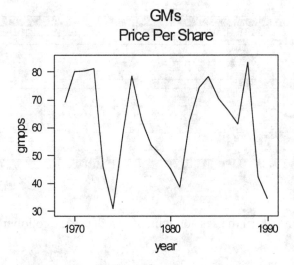

From this plot we can quickly see that GM's price per share varies wildly for the years 1969 to 1990.

What if we want to compare earnings per share and price per share? The best thing to do would be to put earnings per share and price per share on the same graph. The session commands are shown below for graphing "gmeps" and "gmpps" on the same graph.

```
MTB > Plot 'gmeps'*'year' 'gmpps'*'year';
SUBC>    Connect;
SUBC>    Title "GM's";
SUBC>    Title "Earnings Per Share and Price Per Share";
SUBC>    Overlay;
SUBC>    Axis 2;
SUBC>      Label "gmeps,gmpps".
```

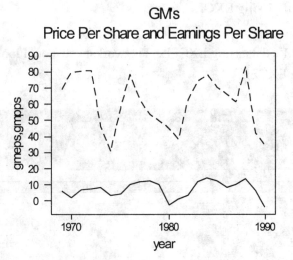

GM's
Price Per Share and Earnings Per Share

In the above graph the dashed line is the price per share and the solid line is the earnings per share.

We can quickly glance at this graph and see that as price per share increases, earnings per share also increases. As price per share decreases, earnings per share decreases. It is also clear that price per share is generally a lot higher than earnings per share. The benefit of this graph is it is easy to understand the *conceptual* relationship between price per share and earnings per share. The disadvantage is that we do not know with precision the relationship between price per share and earnings per share. This problem will be solved when we do statistical calculations so we can precisely compare the two data sets. This simple example shows that graphs are a great way to get quickly a conceptual understanding of a data set but they are not very good at stating precise relationships.

In the next chapter we will look at additional graphical methods for understanding a data set. In Chapter 4 we will look at methods for the precise comparison of data sets.

2.3 Analyzing GM's Data Using EXCEL

We will now analyze GM's data using EXCEL. The worksheet should look like the following after entering the data in EXCEL.

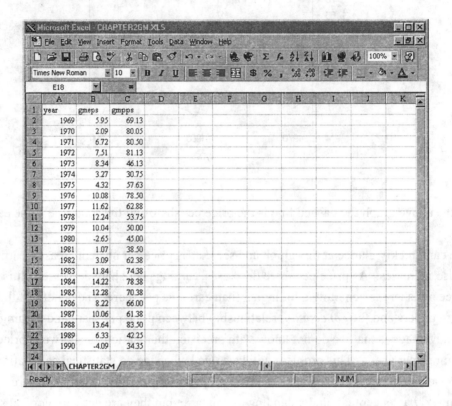

The first thing we need to do is to highlight the data we are interested in charting. One way to do this is to press the *F5* key and type "a1:c23" in the *Reference* option in the *Go To* dialog box as shown below.

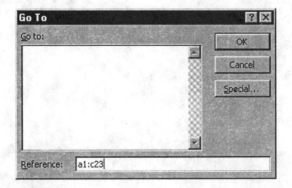

DATA COLLECTION AND PRESENTATION   17

Your worksheet should now look like the following after you press the *OK* button.

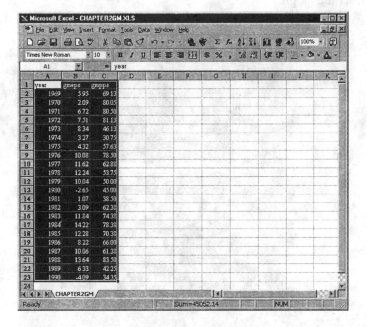

Let us start graphing GM's data by choosing **Insert → Chart**. This will show the following *Chart Wizard* dialog box. Since we are interested in creating a line chart, choose the line option in the *Chart Type* option and in the *Chart sub-type* choose the one shown below.

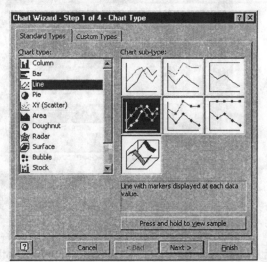

Press the *Next* button to set to step 2 of the chart wizard. The *Data Range* tab of the chart wizard shows the data range that EXCEL will plot. This tab also has a preview of what the chart

18   CHAPTER 2

is going to look like. The preview of the GM's data is wrong. We are going to have to edit the *Series* tab to correct the graph.

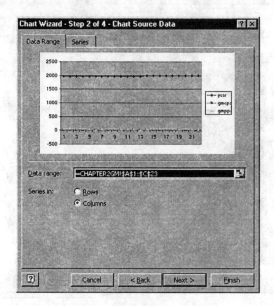

The *Series* option in the *Series Tab* indicates the y-axis and x-axis data range.

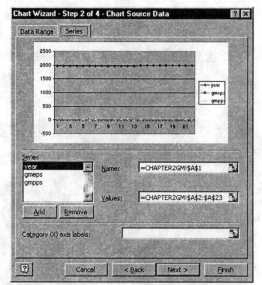

Since we are interested in charting *gmeps* we need to delete the other data. Do this by highlighting the other data and then clicking on the *Remove* button.

DATA COLLECTION AND PRESENTATION   19

We now need to indicate the data range for the x-axis. Put the cursor in *Category (x) axis Labels options* as shown below.

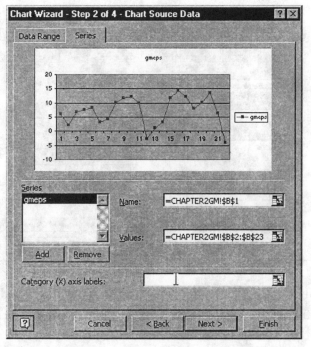

Then highlight the year data as shown below.

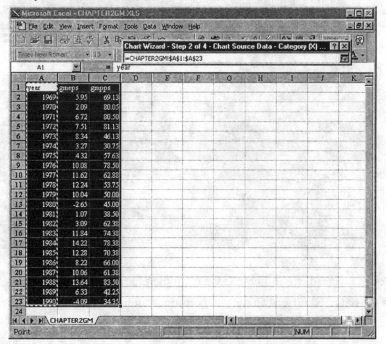

Then press *Enter* on your keyboard to return to the chart wizard. Press *Next* button to go to step 3 of the chart wizard. Fill out step 3 of the wizard as shown below.

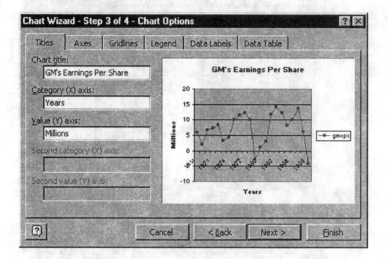

Since we are only showing one data series, let us not show the legend. To do this click on the *Legend* tab and fill out the *Legend* tab as shown below. When finished click on the *Next* button.

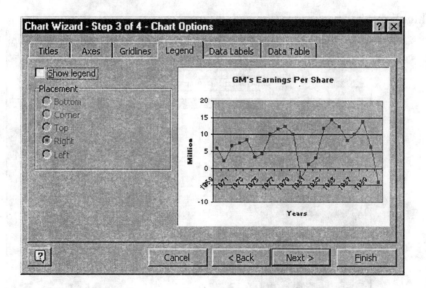

In step 4 of the wizard, we indicate if we want the chart on its own separate sheet or on the same sheet as the data. Choose *As new sheet* option to indicate that we want the chart on a separate sheet. Then we click on the *Finish* button to finish the chart.

DATA COLLECTION AND PRESENTATION   21

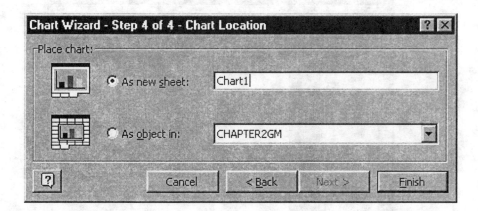

Excel then displays the chart as follows.

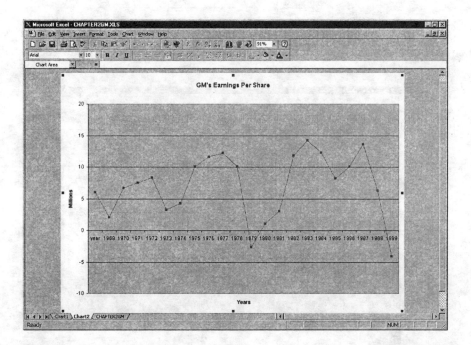

What if we want to compare earnings per share and price per share? The best thing to do would be to put earnings per share and price per share on the same chart. To do this we would highlight the earnings per share data as shown below and then choose **Edit → Copy.**

22   CHAPTER 2

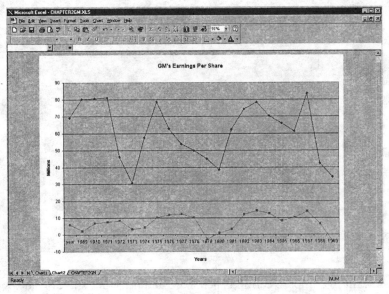

Then make the chart active by choosing it. Then choose **Edit → Paste** to get the following graph.

Choose the chart wizard button as shown below to format the chart further if necessary.

DATA COLLECTION AND PRESENTATION 23

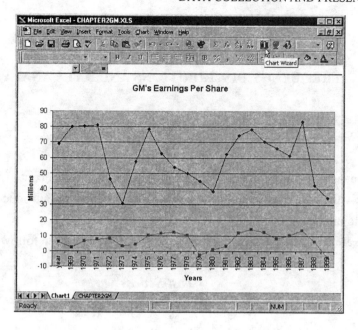

2.4 Statistical Summary

In this chapter we analyzed GM's annual earnings per share and price per share by graphing them. We saw that graphing a data set was a great way to understand it conceptually. But graphs were not very good at describing a data set precisely.

CHAPTER 3
FREQUENCY DISTRIBUTIONS AND DATA ANALYSES

3.1 Introduction

As we saw in the last chapter, graphing the data gives us a general concept of the data more easily and quickly than looking at data in a table. In this chapter we introduce additional graphical ways to understand data sets.

Suppose we are interested in understanding the significance of the following exam scores.

Test Scores

87 83 84 75 86 86 62 95 83 98 87
76 84 83 71 72 65 83 69 64 84 83

By just looking at these test scores, it is hard to understand them. One simple approach would be to tally scores -- that is, see how often certain numbers occur. To do this let us put all the data in column c1 of a MINITAB worksheet. Enter the data from left to right. Name column c1 "grades". The worksheet should look like the following after you have entered all 22 exam scores.

	C1 grades	C2	C3	C4	C5	C6	C7
1	87						
2	83						
3	84						
4	75						
5	86						
6	86						
7	62						
8	95						
9	83						
10	98						
11	87						
12	76						
13	84						
14	83						
15	71						
16	72						
17	65						
18	83						
19	69						
20	64						
21	84						
22	83						
23							
24							

3.2 Tally in MINITAB

We will use the *tally* command to tally the data by typing *tally* and then the column or column name that we want to tally. The *tally* command can analyze only integer data. The following are the session commands to tally up the grades.

```
MTB > tally c1;
SUBC> counts;
SUBC> cumcounts;
SUBC> percents;
SUBC> cumpercents.
```

Summary Statistics for Discrete Variables

grades	Count	CumCnt	Percent	CumPct
62	1	1	4.55	4.55
64	1	2	4.55	9.09
65	1	3	4.55	13.64
69	1	4	4.55	18.18
71	1	5	4.55	22.73
72	1	6	4.55	27.27
75	1	7	4.55	31.82
76	1	8	4.55	36.36
83	5	13	22.73	59.09
84	3	16	13.64	72.73
86	2	18	9.09	81.82
87	2	20	9.09	90.91
95	1	21	4.55	95.45
98	1	22	4.55	100.00
N=	22			

The *tally* command prints a one-way table for each column. The semicolon indicates that a subcommand will follow. The *counts* subcommand counts the frequency of occurrence of each score in the data set. The *cumcounts* subcommand calculates the cumulative frequency of the scores. The *percents* subcommand calculates the percentage of each score in the data set. The *cumpercents* subcommand calculates the cumulative percentage of the scores.

The result tells us that a score of 83 was the most frequent, accounting for nearly 23 percent of all the scores. The interval 83 to 87 had the most occurrences greater than one. These insights would have been hard to derive from the original list of scores.

26 CHAPTER 3

3.3 Tally in Excel

We will now do an equivalent *Tally* analysis in EXCEL. When using the *tally* command in MINITAB, MINITAB produced an analysis for every grade. In EXCEL we could also do this, but we will not. Instead we will analyze grade ranges.

In cell a1 type in "grades". Below cell a1 enter the grades. In cell type b1 "bin". Below cell b1 enter the following 60,65,70,75,80,85,90,100. The EXCEL worksheet should look like the following,

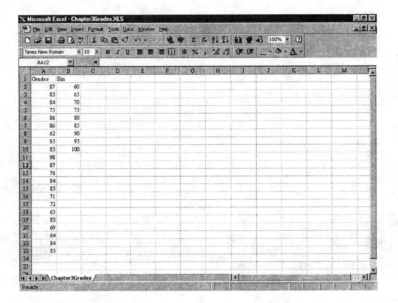

Now choose **Tools → Data Analysis** to get the *Data Analysis* dialog box.

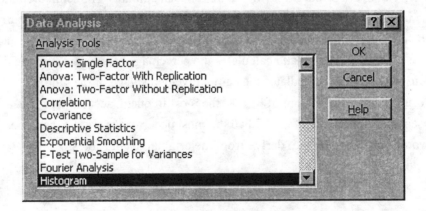

FREQUENCY DISTRIBUTIONS AND DATA ANALYSES 27

In the *Data Analysis* dialog box, choose *histogram* to get the following.

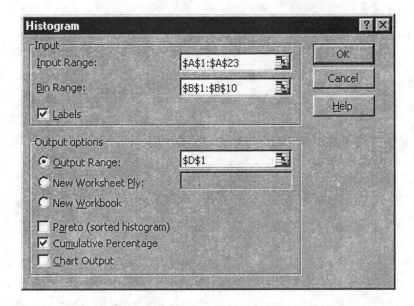

Complete the *Histogram* as shown above. The *Labels* options indicate that the first element in the range indicates the label for the data.

After pressing the *OK* button the following report is produce in cell d1.

Bin	Frequency	Cumulative %
60	0	.00%
65	3	13.64%
70	1	18.18%
75	3	31.82%
80	1	36.36%
85	8	72.73%
90	4	90.91%
95	1	95.45%
100	1	100.00%
More	0	100.00%

3.4 Dotplot

We can look at the exam score graphically by using the *dotplot* command. To use the *dotplot* command, type *dotplot* and then the column or column name that contains the data.

28   CHAPTER 3

```
MTB > dotplot c1
```

Character Dotplot

As seen, the *dotplot* command plots a one-dimensional scatter plot. The first argument ('c1' in this case) specifies the location of the data for the dotplot.

3.5 Histogram

Another way to look at data graphically is by means of histograms. MINITAB uses the *histogram* command to group data into intervals and then graph them. In MINITAB there are two types of graphs, One type of graph is called *Standard Graphs* which are basically graphs produced from characters. The other type is called *Professional Graphs* which are higher resolution graphs. The default graph type is the professional graph. Use the *gstd* session command to create standard graphs. Use the *gpro* session command to create professional graphs. The following is a standard histogram graph.

```
MTB > gstd
MTB > histogram c1;
SUBC> start 65;
SUBC> increment 10.

Histogram of C1    N = 22

Midpoint    Count
    65.0        4    ****
    75.0        4    ****
    85.0       12    ************
    95.0        2    **
```

Notice that the *histogram* command is followed by the column that contains the data and then by a semicolon. The semicolon indicates that we want to issue a subcommand, which is a command that complements the main command. The *histogram* command is a main command. The *start* subcommand indicates that the first midpoint of the histogram is 65. The *increment* subcommand indicates that the histogram will increase by ten. The period after the ten indicates in MINITAB that we are finished with the main command and do not wish to issue any more subcommands.

The following are the session commands to produce a professional graph.

```
MTB > gpro
MTB > Histogram 'grades';
SUBC>   MidPoint;
SUBC>   Bar;
SUBC>   Title "Grade Distribution".
```

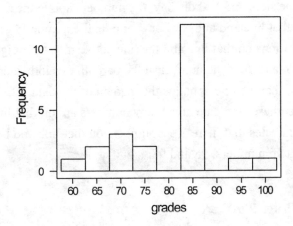

3.6 Stem and Leaf

Another way to show data visually is by using the *stem and leaf* command.

```
MTB > Stem-and-Leaf 'grades';
SUBC>   Increment 10.
```

Character Stem-and-Leaf Display

```
Stem-and-leaf of grades    N = 22
Leaf Unit = 1.0

     4      6  2459
     8      7  1256
   (12)     8  333334446677
     2      9  58
```

30   CHAPTER 3

The *stem and leaf* command produces a stem and leaf graph. The first argument, 'grades', is the location of the data to be analyzed. The subcommand *increment* establishes an interval between the lines. The argument, '10', of the *increment* subcommand sets up an increment of 10 between lines.

The above stem and leaf analysis looks very much like the histogram, and it gives the same information. Both the stem and leaf and the histogram indicate that there are 4 numbers in the 60s interval. The advantage of the stem and leaf, however, is that we know the exact numbers in each interval. From the stem and leaf graph we can see that the 60s interval contains the numbers 62, 64, 65, 69. This kind of information could not be obtained from the histogram.

In the first column of the stem and leaf display, the numbers above the parentheses indicate how many numbers are on that line and above. For example, the number 8 in the second line says that there are 8 observations on that line and the line above it. These eight numbers would be 62, 64, 65, 69, 71, 72, 75, and 76. The number in the parentheses indicates how many numbers are in that line and that the line contains the median of the data set. The numbers in the first column below the parentheses indicate how many numbers are on that line and below. The number 2 on the last line indicates that there are 2 numbers on that line and below. Since this is the last line, there are only two numbers, 95 and 98.

3.7 Boxplot (Box-and-Wisker Plot)

```
MTB > GStd.
MTB > BoxPlot 'grades'.
```

Character Boxplot

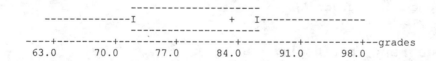

The *boxplot* command will boxplot the data in column c1. The + in the boxplot is the middle observation of a data set. In our grades data set it would be 83. This number, called the median, will be discussed in Section 4.5 in the next chapter. The box contains the middle 50% of the grades data. To the left of the box are 25% of the data and to the right of the box are the remaining 25% of the data. In other words, the box-and-whisker plot is a graphical presentation of a set of sample data that illustrates the first 25% (first quartile), the second 25% (second quartile), the third 25 % (third quartile), and the fourth 25% (fourth quartile) of a sample data set. Note that the end of the second 25% or second quartile of a data set is the median. Using MINITAB to calculate all these statistics will be discussed in Section 4.5

Below shows the session commands for the professional boxplot graph.

```
MTB > gpro
MTB > Boxplot 'grades';
SUBC>    Transpose.
```

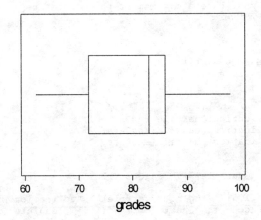

We will now create a boxplot in EXCEL. Since there is no boxplot feature in EXCEL we will write an program to create a boxplot. The program below is the EXCEL program to create a boxplot. The code below is complicated and it is beyond the scope of this book to discuss how to program in EXCEL.

Look at the next section on where to put program code in EXCEL and how to execute program code in EXCEL.

CHAPTER 3

```vba
'/***********************************************************
'/Purpose:   Draws a boxplot based on the data in column a
'/          Put all data in column a
'/***********************************************************
Sub DrawBoxPlot()
    Dim sngX(0 To 13) As Single
    Dim sngY(0 To 13) As Single
    Dim sngLower_inner_fence As Single
    Dim sngUpper_inner_fence As Single
    Dim sngInterQuartile
    Dim sngQuartile(4) As Variant
    Dim rngData As Range
    Dim wkCallingSheet As Worksheet
    Dim sngOutlierX() As Single
    Dim sngOutlierY() As Single
    Dim nCounter As Integer
    Dim rngNumber As Range


    Set rngData = ActiveSheet.Range("a1").CurrentRegion.Columns(1)
    If rngData.Rows.Count < 3 Then
        Call MsgBox("You must have a minimum of three data items in _
            column A!", , "Boxplot")
        Exit Sub
    End If

    Application.ScreenUpdating = False
    ActiveSheet.ChartObjects.Delete


    'Calcualte Quartiles
    sngQuartile(1) = Application.Quartile(rngData, 1)
    sngQuartile(2) = Application.Quartile(rngData, 2)
    sngQuartile(3) = Application.Quartile(rngData, 3)


    sngInterQuartile = sngQuartile(3) - sngQuartile(1)

    If (sngQuartile(1) - (1.5) * sngInterQuartile) > Application.Min(rngData) Then
        sngLower_inner_fence = (sngQuartile(1) - (1.5) * sngInterQuartile)
    Else
        sngLower_inner_fence = Application.Min(rngData)
    End If


    If (sngQuartile(3) + (1.5) * sngInterQuartile) > Application.Max(rngData) Then
        sngUpper_inner_fence = Application.Max(rngData)
    Else
        sngUpper_inner_fence = (sngQuartile(3) + (1.5) * sngInterQuartile)
    End If

    sngX(0) = sngQuartile(1)
    sngY(0) = 1
    sngX(1) = sngQuartile(2)
    sngY(1) = 1
    sngX(2) = sngQuartile(3)
    sngY(2) = 1
    sngX(3) = sngQuartile(3)
    sngY(3) = 1.5
    sngX(4) = sngUpper_inner_fence
    sngY(4) = 1.5
    sngX(5) = sngQuartile(3)
    sngY(5) = 1.5
    sngX(6) = sngQuartile(3)
    sngY(6) = 2
    sngX(7) = sngQuartile(2)
    sngY(7) = 2
```

```
        sngX(8) = sngQuartile(2)
        sngY(8) = 1
        sngX(9) = sngQuartile(2)
        sngY(9) = 2
        sngX(10) = sngQuartile(1)
        sngY(10) = 2
        sngX(11) = sngQuartile(1)
        sngY(11) = 1
        sngX(12) = sngQuartile(1)
        sngY(12) = 1.5
        sngX(13) = sngLower_inner_fence
        sngY(13) = 1.5

        Set wkCallingSheet = ActiveSheet
        Charts.Add
        With ActiveChart
                .ChartType = xlXYScatterLinesNoMarkers
                .SeriesCollection.NewSeries
                .SeriesCollection(1).xValues = sngX
                .SeriesCollection(1).values = sngY
                With .Axes(xlCategory)
                        .HasMajorGridlines = False
                        .HasMinorGridlines = False
                End With
                With .Axes(xlValue)
                        .HasMajorGridlines = False
                        .HasMinorGridlines = False
                        .Delete
                End With
                On Error Resume Next
                .SeriesCollection(2).Delete
                Err = 0
                On Error GoTo 0

                .PlotArea.Interior.ColorIndex = xlNone
                .HasLegend = False
                .Location Where:=xlLocationAsObject, Name:=wkCallingSheet.Name

        End With


        nCounter = 0
        For Each rngNumber In rngData.Cells
            If (rngNumber.Value >= sngUpper_inner_fence) Or (rngNumber.Value _
                <= sngLower_inner_fence) Then
                ReDim Preserve sngOutlierX(nCounter)
                ReDim Preserve sngOutlierY(nCounter)
                sngOutlierX(nCounter) = rngNumber.Value
                sngOutlierY(nCounter) = 1.5
                nCounter = nCounter + 1
            End If
        Next

        If nCounter = 1 Then
            Exit Sub
        Else
            With ActiveChart
                .SeriesCollection.NewSeries
                With .SeriesCollection(2)
                        .xValues = sngOutlierX
                        .values = sngOutlierY
                        .MarkerBackgroundColorIndex = xlAutomatic
                        .MarkerForegroundColorIndex = xlAutomatic
                        .Border.LineStyle = xlNone
                        .MarkerStyle = xlSquare
                        .MarkerSize = 5
```

34   CHAPTER 3

```
                    .Name = "Outliers"
             End With
             .SeriesCollection(1).Name = "BoxPlot"
             .HasLegend = True
             '.PlotArea.Interior.ColorIndex = xlNone
        End With
    End If
    rngData.Range("a1").Select
End Sub
```

Before running the above EXCEL boxplot macro, put the grades data in column A. The boxplot macro produces the following boxplot chart of the grades.

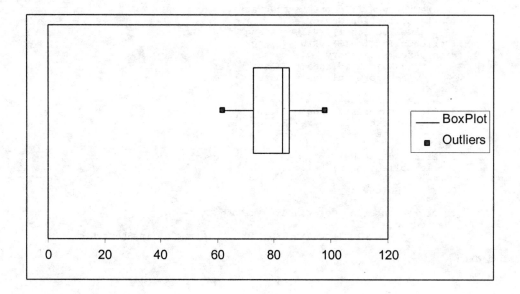

3.8 EXCEL Programming

EXCEL programs are put in *modules*. You use the *visual basic editor* to create and modify modules. To get to the visual basic editor choose **Tools → Macro → Visual Basic Editor** or by pressing the *Alt-F11* keys. Below is the visual basic editor.

FREQUENCY DISTRIBUTIONS AND DATA ANALYSES   35

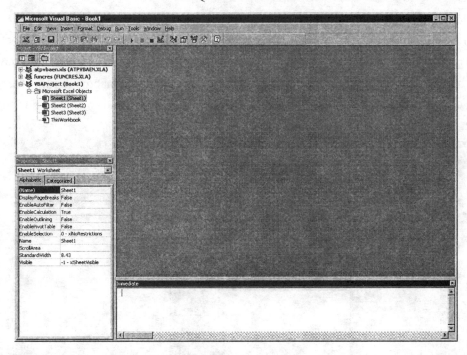

Choose **Insert → Module** to insert a module sheet into the visual basic editor. Below shows the module sheet in the visual basic editor.

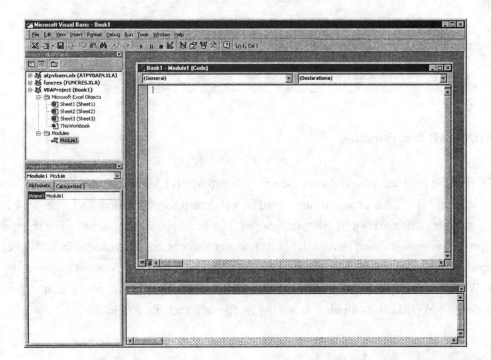

36   CHAPTER 3

Enter the boxplot code into the module sheet. After entering the code press *Alt-F11* to get back to EXCEL.

To run the boxplot code in EXCEL press the *Alt-F8* keys to get the *Macros* dialog box as shown below. In the macros the dialog box shows all the available macros to run. To run a macro, choose the macro to run and press the *Run* button. The boxplot was shown in the last section.

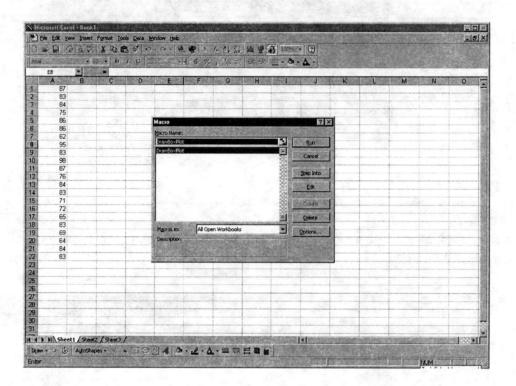

3.9 MINITAB Programming

In EXCEL the program is saved with the data file. In MINITAB the program is saved in a separate file. MINITAB programs are saved in a standard format called *Text Files*. Text files are simple files that are read by most programs. The best type of programs to use text files are wordprocessing programs. Two of the most popular choices are Microsoft Word (WORD) and Notepad. Word comes with Microsoft Office and Notepad with the Windows operating system.

In WORD it is important to make sure that the file is saved as a *text file* format. The *Save* Dialogbox of WORD shown below is saving its file in a text file format.

FREQUENCY DISTRIBUTIONS A...

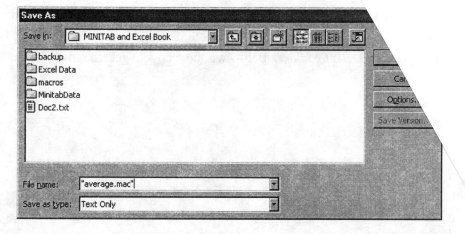

Notice that in the *File name* option has "average.mac". It is important to have the macro name in double quotes and to have a *mac* file extension. MINITAB recognizes *mac* files to be MINITAB program files. The double quotes is to force WORD to accept *mac* as a file extension. If the double quotes where not there, WORD would make the file extension as *txt* files.

The *Save* dialog box of Notepad shown below is saving its file in a text file format.

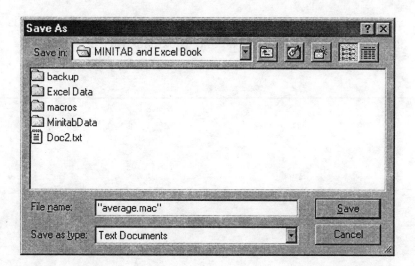

When MINITAB is open it defaults to the *data* directory of MINITAB. The "cd" MINITAB commands below shows the current directory of MINITAB

```
MTB > cd
C:\Program Files\MTBWIN\Data
```

This is important because MINITAB will look at the current directory to execute commands. It does not make sense to save MINITAB programs in a folder called *data*. Instead there is a folder called *macros* under the *MTBWIN* folder. Below we show how to get from the *data* folder to the *macros* folder.

```
MTB > cd
C:\Program Files\MTBWIN\Data

MTB > cd..
MTB > cd
C:\Program Files\MTBWIN

MTB > cd macros
MTB > cd
C:\Program Files\MTBWIN\macros
```

All macros created should be stored in the *macros* folder.

3.10 Statistical Summary

In this chapter we looked at more ways to analyze data graphically. These methods were: tally, dotplot, histogram, stem and leaf, and boxplot (box-and-whisker).

CHAPTER 4
NUMERICAL SUMMARY MEASURES

4.1 Introduction

In the last chapter we looked at graphical methods of analyzing and understanding a data set. In this chapter we look at mathematical methods of analyzing and understanding a data set. We should realize that we get different benefits from graphing data than from analyzing them mathematically. The benefit of graphing a data set is that we can easily understand it conceptually. Mathematical analysis, however, gives a more precise description and is useful for comparing two data sets with each other. Whenever possible, a data set should be graphed and should be mathematically analyzed as well.

4.2 Arithmetic Mean

The arithmetic mean is a measure of central tendency or location. It is a very important statistical concept because many complicated statistical formulas contain the mean. Mathematically it is defined as

$$\bar{x} = \frac{\sum_{i=1}^{n} x_i}{n}$$

In MINITAB it is very easy to get the mean value of a data set. Let us use the set of exam scores that we used in Chapter 3.

```
MTB > names c1 'grades'
MTB > set 'grades'
DATA> 87 83 84 75 86 86 62 95 83 98 87
DATA> 76 84 83 71 72 65 83 69 64 84 83
DATA> end
MTB > mean 'grades'
```

Column Mean

```
    Mean of grades = 80.000
```

From the MINITAB output shown above, the mean exam score is 80.

To calculate the mean in EXCEL, we will use the function wizard. To get to the function wizard, choose **Insert -> Function**.

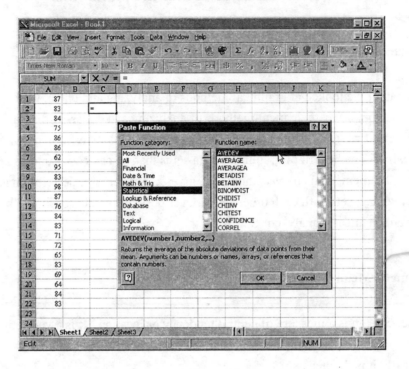

In the *Function category* choose statistical and in the *Function name* choose average then press the *OK* button.

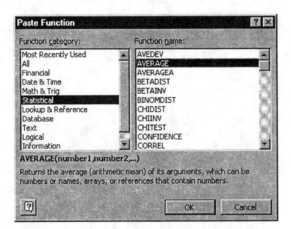

NUMERICAL SUMMARY MEASURES   41

The following dialog box will appear to indicate and explain the parameters for the average function.

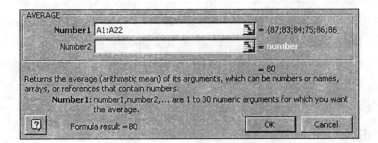

In the first argument type in or highlight range A1:A2 and then press the *OK* button.

As shown above, the average is calculated in cell c2. Notice above that the cursor is at cell c2 and the *formula bar* shows the formula used in cell c2.

4.3 Median and Quartiles

Another measure of the central tendency of a data set is the median. The median is the middle observation of a data set, and we can find it very easily by using the *median* command. The argument of the *median* command is the column or name of the column that contains the data set. Below is the median score for the exam data that we used in Chapter 3.

```
MTB > median 'grades'
```

Column Median

```
Median of grades = 83.000
```

EXCEL has a built-in *median* function. The median calculation of the grades in EXCEL is shown below.

We can present the median graphically by using the box-and-whisker plot as shown in the last chapter.

If we want to divide the data into four quarters, then we can calculate quartile statistics: first quartile (Q1), median (Q2), and third quartile (Q3) can be found using MINITAB commands.

The *median* command is shown above. The following display shows how we can use MINITAB to find Q1, Q2, and Q3 for the exam score data that we used in Chapter 3.

We will show two approach as, a numerical approach and a graphical approach. The numerical approach is by using *describe* command.

```
MTB > describe 'grades'
```

Descriptive Statistics

```
Variable        N       Mean    Median   Tr Mean    StDev   SE Mean
grades         22       80.00    83.00     80.00     9.57      2.04

Variable      Min        Max        Q1        Q3
grades      62.00      98.00     71.75     86.00
```

From the MINITAB output we can see that Q1 is 71.75, Q2 is 83 and Q3 is 86.

The graphical approach is used by the *boxplot* command.

```
MTB > gstd
* NOTE  * Standard Graphics are enabled.
          Professional Graphics are disabled.
          Use the GPRO command to enable Professional Graphics.
MTB > boxplot 'grades'
```

Boxplot

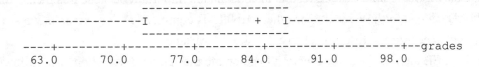

From the boxplot output Q1 is the left side of the box, Q2 is from the left side of the box to the plus sign, Q3 is from the right side of the box to the plus and Q4 is to the right of the box.

This is a good example where by using a numerical method we can precisely know the quartile numbers and by using a graphical method we can conceptually understand the quartile numbers.

4.4 Standard Deviation

The standard deviation is a measure of the spread of a data set. The mathematical formula for the standard deviation is

$$s = \sqrt{\frac{\sum_{i=1}^{n}(x_i - \bar{x})^2}{n-1}}$$

The *stdev* MINITAB command is used to calculate the standard deviation. It is illustrated below.

```
MTB > stdev 'grades' k1
```

Column Standard Deviation

```
Standard deviation of grades = 9.5718
```

The above stdev command calculates the standard deviation for grades and puts the result in k1.

k1, called a constant, is a storage location in MINITAB that cannot be viewed anywhere on the data sheet. Being able to store results in constants is important because this means that MINITAB can get to these results for other MINITAB commands. For example, we are going to have to get access to the result in k1 to calculate the variance.

In EXCEL there is the *stdev* function to calculate the standard deviation. The following shows the calculation of the standard deviation of the data set.

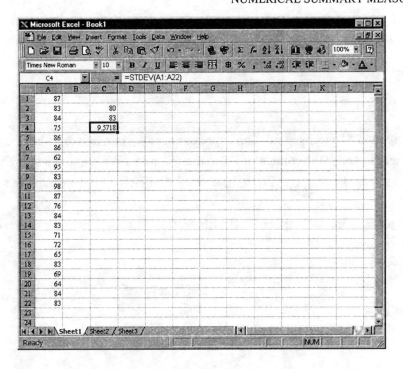

4.5 Variance

The variance is another measure of the spread, or dispersion, of a data set. Mathematically it looks like the following.

$$\sigma^2 = \sqrt{\frac{\sum_{i=1}^{n}(x_i-\bar{x})^2}{n-1}} * \sqrt{\frac{\sum_{i=1}^{n}(x_i-\bar{x})^2}{n-1}} = \frac{\sum_{i=1}^{n}(x_i-\bar{x})^2}{n-1}$$

From this definition we can see that the variance is the square of the standard deviation. This is a useful observation, because in MINITAB there is no command for the variance.

In the previous section we stored the standard deviation in the constant k1. We will square the constant k to get the variance. The variance will be stored in k2 This is shown below.

46   CHAPTER 4

```
MTB > let k2 = k1 ** 2
MTB > print k2
```

Data Display

```
K2    91.6190
```

The expression ** means to raise an expression to a certain power. The **2 in the *let* command means to raise the content of K2 to the second power. In this way we tell MINITAB to square the standard deviation. From the above MINITAB output we can see that the variance is 91.6190.

How do we interpret the variance? To illustrate, let usok at another data set of exam scores with a smaller variance.

Test Scores Set 2

87	83	84	98	86	86	91	95	83	98
91	84	83	94	83	86	83	97	93	84
83	87								

To calculate the variance of these scores, we put them into column c2 and then issue the following commands:

```
MTB > set c2
DATA> 87 83 84 98 86 86 91 95 83 98 91
DATA> 84 83 94 83 86 83 97 93 84 83 87
DATA> end
MTB > stdev c2 k3
```

Column Standard Deviation

```
   Standard deviation of C2 = 5.4362

MTB > let k4 = k3**2
MTB > print k4
```

Data Display

```
K4    29.5519
```

The result indicates that data set 2 has a smaller variance than data set 1. This means that data set 2 has less spread among its data. We will plot the two sets of test scores to confirm this.

Exam Scores - Data Set 1

```
MTB > dotplot c1;
SUBC> start 60;
SUBC> increment 10.
```

Dotplot

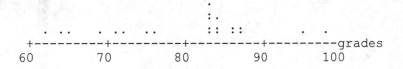

Exam Scores - Data Set 2

```
MTB > dotplot c2;
SUBC> start 60;
SUBC> increment 10.
```

Dotplot

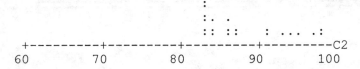

We can see visually that the spread in data set 2 is smaller than in data set 1. This agrees with the variance and standard deviation numbers that we calculated for both of these data sets. It is important to note that the standard deviation and the variance represent the spread of the same data set even though they are different numbers. For example, the standard deviation and the variance for data set 2 are 5.4363 and 29.552, respectively. This is like explaining data set 2 in two different languages.

Keep in mind that we cannot judge the magnitude of a variance or standard deviation until we compare it with another variance or standard deviation. We cannot say whether the variance of the second data set, 29.552, is small until we compare it with the variance of the first data set, which is 91.619. Not until we compare the second data set variance to the first data set variance, do we realize that the second variance is small compared to the first.

48   CHAPTER 4

The variance and the standard deviation, like the mean, are very important statistical concepts. Other important statistical calculations will depend on them.

In EXCEL there is the *var* function to calculate the variance. The following shows the calculation of the variance of the data set.

4.6 *Z* Score

The *Z* score expresses the relative position of any particular data item in terms of the number of standard deviations above or below the mean. The score is a very important concept that will be used extensively in future chapters. It is defined as

$$Z = \frac{x - \bar{x}}{s}$$

A zero *Z* score indicates that a particular data value is equal to the mean. A positive 1 *Z* score indicates that a data value is one standard deviation to the right of the mean. A negative 1 *Z* score indicates that a data value is one standard deviation to the left of the mean.

A very important feature of mound-shaped data sets is that approximately 68 % of the observations have a Z score between –1 and 1, and approximately 95 % of the observations have a Z score between –2 and 2.

This is graphically shown below.

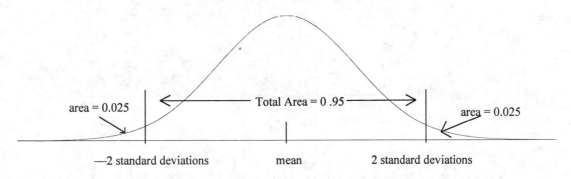

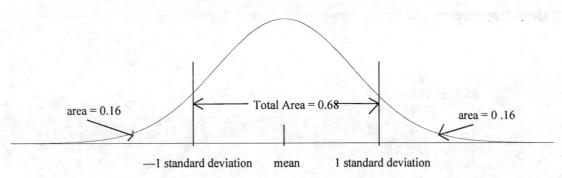

We will use the *center* session command to calculate the Z score for every value in a data set.

Example 4.6

We will calculate the Z score of the following data set

20	42	56	78
21	43	58	80
23	43	59	81
25	46	61	85
30	48	62	90
35	50	65	92
36	51	68	96
39	52	70	98
40	54	71	99
41	55	75	100

```
MTB > set c1
DATA> 20 21 23 25 30 35 36 39 40 41
DATA> 42 43 43 46 48 50 51 52 54 55
DATA> 56 58 59 61 62 65 68 70 71 75
DATA> 78 80 81 85 90 92 96 98 99 100
DATA> end
MTB > center c1 c2
MTB > print c1 c2
```

Data Display

Row	C1	C2
1	20	-1.67276
2	21	-1.62925
3	23	-1.54224
4	25	-1.45523
5	30	-1.23771
6	35	-1.02019
7	36	-0.97668
8	39	-0.84617
9	40	-0.80266
10	41	-0.75916
11	42	-0.71565
12	43	-0.67215
13	43	-0.67215
14	46	-0.54163
15	48	-0.45462
16	50	-0.36761
17	51	-0.32411
18	52	-0.28061

```
19     54    -0.19360
20     55    -0.15009
21     56    -0.10659
22     58    -0.01958
23     59     0.02393
24     61     0.11094
25     62     0.15444
26     65     0.28496
27     68     0.41547
28     70     0.50248
29     71     0.54598
30     75     0.72000
31     78     0.85052
32     80     0.93753
33     81     0.98103
34     85     1.15505
35     90     1.37257
36     92     1.45958
37     96     1.63360
38     98     1.72061
39     99     1.76412
40    100     1.80762
```

4.7 Skewness

Sometimes we need to know whether a data set exhibits a symmetric pattern. The *coefficient of skewness* answers this question. The formula for the coefficient of skewness is

$$CS = \frac{\sum_{i=1}^{N}(x_i - \mu)^3}{\sigma^3}$$

We interpret skew numbers as follows:

Zero skewness coefficient means that the distribution is symmetric, with the mean equal to the median.

Positive skewness coefficient means that the distribution is skewed to the right and that the median lies below or to the left of the mean.

Negative skewness coefficient means that the distribution is skewed to the left and the median is above or to the right of the mean.

These concepts are graphically illustrated below.

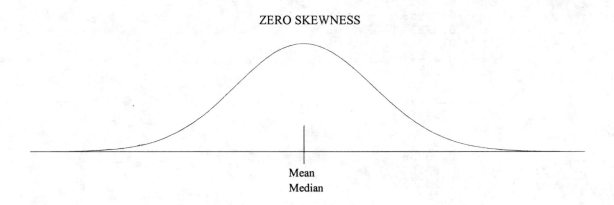

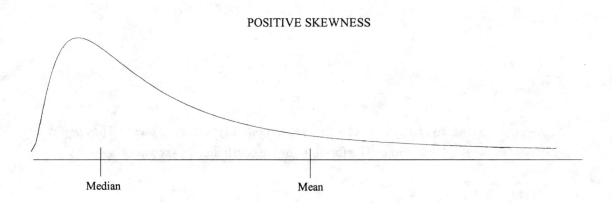

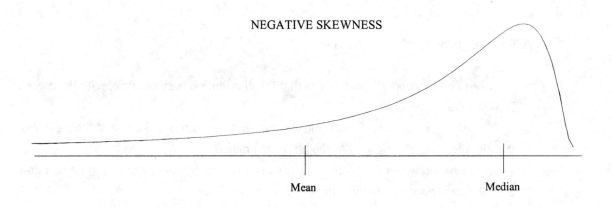

Notice that for the zero skewness, the hump of the distribution is in the middle. The hump of the positive skewness is towards the left. The hump of the negative skewness is towards the right.

Example 4.7A

Suppose we are interested in the skewness of the following numbers,

$$3, 5, 8, 14$$

To calculate the skewness let us put the above numbers in column c1. To calculate the skewness command we will use the *Stat* session command and the *skewness* subcommand. This is shown below.

```
MTB > set c1
DATA> 3 5 8 14
DATA> end
MTB > stat c1;
SUBC> skewness c2.
MTB > print c2
```

Data Display

```
C2
   1.01537
```

The output above tells us that the numbers are skewed positively, so the median is to the left of the mean.

Example 4.7B

Suppose we are interested in the skewness of a group of ACT scores. The data show that

8 people got a score of 16
34 people got a score of 21
20 people got a score of 26
10 people got a score of 30
8 people got a score of 36

We can enter the data into the worksheet individually or we can use the *set* command. The advantage of using the *set* command is that we can enter data in sequences. Entering the above eighty data items using the *set* command is illustrated below.

```
MTB > set c1
DATA> 8(16)
DATA> 34(21)
DATA> 20(26)
DATA> 10(31)
DATA> 8(36)
DATA> end
MTB > stat c1;
SUBC> skewness c2.
MTB > print c2
```

Data Display

```
C2
   0.597383
```

From the output we see that the ACT test score is positively skewed and the median is to the left of the mean.

4.8 Statistical Summary

In this chapter we looked at various methods of describing a data set. We can now group these measurements as follows.

1. Measure of Central Tendency or Measure of Location
    a. arithmetic mean
    b. median
    c. geometric mean
This group tells us the central location of a data set.

2. Measure of Spread.
    a. standard deviation
    b. variance
This group tells us how spread out the data set is.

3. Measure of Relative Position
    a. quartiles
    b. Z score

This group tells us the relative position of a data value in a data set.

4. Measure of Shape
    a. Skewness

This measure tells us if a data set is symmetrical relative to the mean or median.

CHAPTER 5
PROBABILITY CONCEPTS AND THEIR ANALYSIS

5.1 Introduction

In the previous chapters we looked at descriptive statistics. Now we will begin the study of inferential statistics and show how MINITAB and EXCEL can help. Inferential statistics analyzes a partial data set and, on the basis of the analysis, makes educated inferences about the complete data set. In the process, inferential statistics makes uses of descriptive statistics and probability. It is for this reason that we study probability concepts in this chapter. The partial data set is usually called a *sample*. The complete data set is usually called a *population*.

Below is a graphical view of the relationships among sample data, population data, descriptive statistics, inferential statistics, math, probability, and MINITAB and EXCEL.

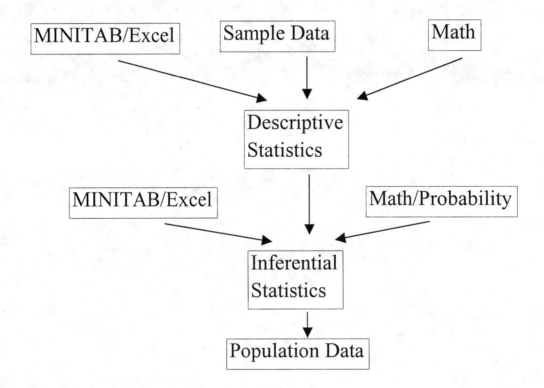

5.2 Probability

One of the most fundamental concepts in probability is that the probability of the total number of possible outcomes equals 100% (many times referred to as 1) and the probability of an item or items from the total number of possibilities is between 0% and 100% (that is, between 0 and 1).

As a corollary, the sum of the probability of all the items of the total number of possible outcomes equals 1.

Suppose there are only two items, A and B. Then

$$\text{probability (A)} + \text{probability (B)} = 1$$

Therefore, if we know the probability of B, we should know the probability of A by rearranging the above formula to

$$\text{probability (A)} = 1 - \text{probability (B)}$$

An underlying issue is knowing the total number of possible outcomes and knowing what items we are interested in and finding the percentage of these items compared to the total number of possible outcomes.

$$\text{Probability of an event} = \frac{\text{items of interest}}{\text{total number of possible outcomes}}$$

For example, the total number of possible outcomes of tossing a die is six because there are only six possible numbers. The probability of tossing the number 3 therefore would be $1/6 = 0.17$. If we change to two dice, then the total number of possible outcomes would be twelve. Then, if we ask what is the chance of getting a three when we throw the two dice, the probability would be $2/12 = 0.17$.

Example 5.2A

Below is a roster of a class of ten students which consists of accounting and finance students. What percentage of the class is accounting students? What percentage is female?

Sex	Major
1	2
1	1
2	2
1	1
1	1
2	1
1	2
1	1
1	2
2	1

Sex	Major
1=Female	1=Accounting
2=Male	2=Finance

We can find the percentage of females and the percentage of accounting students by using the *table* command. We will put the sex data in column c1 and name the column c1 '*sex*'. We will put the major data in column c2 and name the column c2 '*major*'.

The *table* command and its effects are illustrated below.

```
MTB > set c1
DATA> 1 1 2 1 1 2 1 1 1 2
DATA> end
MTB > set c2
DATA> 2 1 2 1 1 1 2 1 2 1
DATA> end
MTB > table c1 c2;
SUBC> totprecents.
```

Tabulated Statistics

```
 Rows: C1      Columns: C2
              1         2        All

  1       40.00     30.00      70.00
  2       20.00     10.00      30.00
 All      60.00     40.00     100.00

  Cell Contents --
                % of Tbl
```

In the table, the rows represent Sex and the columns represent Major. All the information is in percentages. We can see from the table that the class consists of 70% females and 60% accounting students. The table also tells us that 40% of the students are female accounting students and 30% are female finance students. Twenty percent are male accounting students and 10% are male finance students.

5.3 Permutations

Suppose we are interested in the number of possible arrangements of the letters A, B, C, and order is important. That is, AB and BA are two different arrangements. The concept of permutations can help us answer our question. The mathematical notation for permutation is

$$_nP_r = \frac{n!}{(n-r)!}$$

n in the permutation formula is the number of elements in the population, and r is the number of elements in the sample.

MINITAB does not contain a command to calculate the permutation, and so we will create our own command, or macro.

Perm Macro

```
gmacro
noecho

#Define Variables
name k50 = 'n'
name k51 = 'r'
name k52 = 'loopcounter'

note What is the n for the permutation?
set 'terminal' c50;
nobs=1.
end
let 'n' = c50

note What is the r for the permutation?
set 'terminal' c50;
nobs=1.
end
let 'r' = c50
let c60=1

do 'loopcounter' = 1:'r'
     let c60=c60*'n'
     let 'n'='n'-1
enddo

note The permutation value is:
name c60= '--------'
print c60
erase c40-c70
endmacro
```

Example 5.3A

Find the number of permutations for the letters A, B, C. To do this we will use the *perm* macro.

```
MTB > %perm
Executing from file: perm.MAC
What is the n for the permutation?
DATA> 3
What is the r for the permutation?
DATA> 3
The permutation value is:

Data Display

--------
    6
```

In this case, since the question is simple, we can confirm that the permutation value is six by listing the six possible arrangements.

ABC	BAC	CAB
ACB	BCA	CBA

Knowing that the permutation value of ABC is six, we know that the probability of getting the specific permutation ABC is 1/6 = 0.17.

The code below is EXCEL code that shows all the permutations when *r* is equal to *n*. The code consists of three procedures and one function

```
Dim CurrentRow As Variant
Dim CurrentColumn As Variant
'    The source of this algorithm is unknown

Sub GetString()
    Dim InString As String
    Dim tstart As Variant

    tstart = Now()

    InString = InputBox("Enter text to permutate:")
    If Len(InString) < 2 Then Exit Sub
    ActiveSheet.Cells.Clear
    CurrentRow = 1
    CurrentColumn = 1
    Call GetPermutation("", InString, tstart)
    Application.StatusBar = False
End Sub

Sub GetPermutation(x As String, y As String, tstart As Variant)

    Dim i As Integer, j As Integer
    j = Len(y)
    If j < 2 Then

        Application.StatusBar = "Processing cells " & CurrentRow & "," & CurrentColumn _
                             & " Permuations Calculated:" & _
                             ((ThisWorkbook.Worksheets.Count - 1) * 16777216) + _
                             (CurrentColumn - 1) * 65536 + CurrentRow _
                             & " Elasped Time(hh:mm:ss): " & Format(tstart - Now(), _
                             "hh:mm:ss")
        Cells(CurrentRow, CurrentColumn) = x & y
        CurrentRow = CurrentRow + 1
        If CurrentRow = 65537 Then
            CurrentRow = 1
            CurrentColumn = CurrentColumn + 1
            If currentcurrentcolumn = 257 Then
                CurrentRow = 1
                CurrentColumn = 1
                ThisWorkbook.Worksheets.Add
            Else
                ActiveSheet.Columns(CurrentColumn - 1).EntireColumn.AutoFit
            End If
        End If
    Else
        For i = 1 To j
            Call GetPermutation(x + Mid(y, i, 1), _
            Left(y, i - 1) + Right(y, j - i), tstart)
        Next
    End If
End Sub

Public Function Time_Elasp(tstart As Variant) As Variant

 Time_Elasp = Format(tstart - Now(), "hh:mm:ss")

End Function
```

The use of the above code to illustrate Example 5.3A is shown below. Execute the *getstring* procedure to get the following.

CHAPTER 5

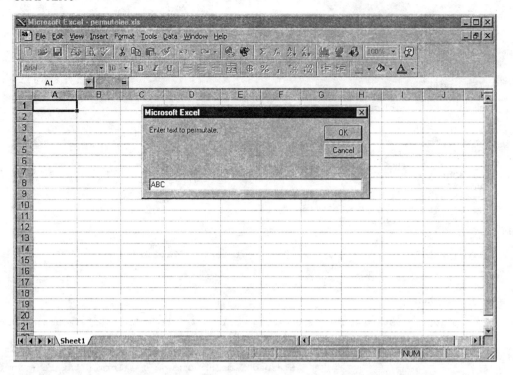

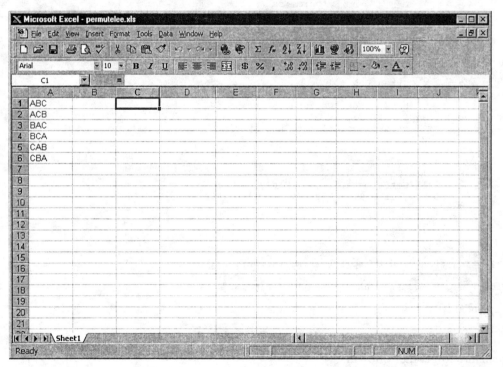

EXCEL also has a built-in function call *permut* to calculate the number of permutations. This is shown below.

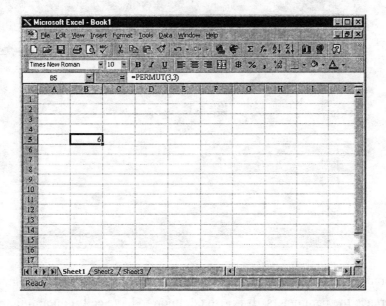

Example 5.3B

Permutation numbers can get very large very quickly. Suppose we are interested in how many ways we can permutate the alphabet in groups (or 'items of interest') of six letters each. This is illustrated below.

```
MTB > %perm
Executing from file: perm.MAC
What is the n for the permutation?
DATA> 26
What is the r for the permutation?
DATA> 6
The permutation value is:
```

Data Display

```
--------
  165765600
```

We can see from the *perm* macro that there are 165,765,600 permutations. To compute probabilities, we often need to calculate permutations. The 'birthday problem' is a popular example of probability based on permutations. We will now calculate the 'birthday problem' in EXCEL. This is shown below.

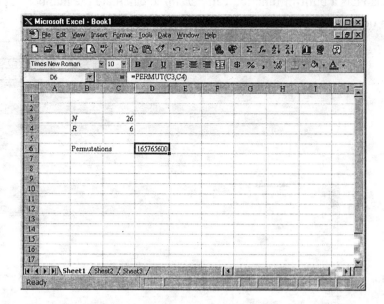

It is common in EXCEL to have cells containing the parameters of the function and have the function reference the parameters. It is also common to have cells containing the value of the parameters have labels to the left of the cell to describe the cell.

5.4 Combination

If we are not interested in order, then we can use the concept of combination to count. In this case, the arrangements AB and BA are equivalent. Mathematically, the concept of counting is as follows:

$$_nC_r = \frac{n!}{r!(n-r)!}$$

As with permutation, we do not have a MINITAB command to calculate the combination. Instead, we will have to create our own combination command or macro. First, however, let us look more closely at the mathematical formulas for both combination and permutation.

$$_nC_r = \frac{n!}{r!(n-r)!} \quad , \quad _nP_r = \frac{n!}{(n-r)!}$$

We can see that there is really only one difference between combination and permutation. That difference is that the combination formula has an *r*! factorial in the denominator and the permutation formula does not. This tells us that construction of combination macros will be very similar to that of permutation macros.

Comb Macro

```
gmacro
noecho
#Define Variables
name k50 = 'n'
name k51 = 'r'
name k52 = 'loopcounter'
note What is the n for the combination?
set 'terminal' c50;
nobs=1.
end
let 'n' = c50
note What is the r for the combination?
set 'terminal' c50;
nobs=1.
end
let 'r' = c50
let c60=1
do 'loopcounter' = 1:'r'
     let c60=(c60*'n')/'r'
     let 'r'='r'-1
     let 'n'='n'-1
enddo

note The combination value is:
name c60= '--------'
print c60
erase c40-c70
endmacro


MTB > %comb
Executing from file: comb.MAC
What is the n for the combination?
DATA> 5
What is the r for the combination?
DATA> 3
The combination value is:
```

Data Display

```
--------
    10
```

Example 5.4A

Suppose we are interested in how many ways we can arrange the letters A, B, C, D, E into groups ('items of interest') of 3 letters each and we are not interested in the order of the letters.

```
MTB > %comb
Executing from file: comb.MAC
What is the n for the combination?
DATA> 5
What is the r for the combination?
DATA> 3
The combination value is:
```

Data Display

```
--------
    10
```

We can confirm this answer by listing the 10 combinations:

```
ABC   BCD   CDE   EAC
ABD   BCE   DEA
ABE   CDA   DAC
```

EXCEL has a built-in function call *combin* to calculate the number of combinations. This is shown below.

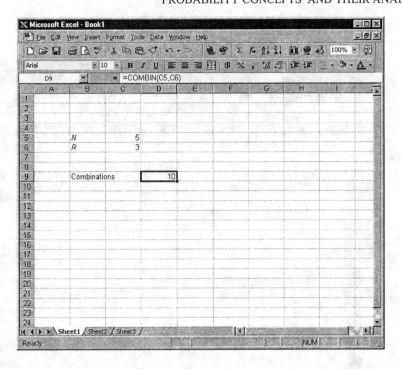

Example 5.4B

Earlier, we looked at the permutation of the alphabet for items of interest with six letters. Let us now look at the combination of the alphabet for items of interest with six letters. The answer should be less, because we are not interested in the order of the items of interest.

```
MTB > %comb
Executing from file: comb.MAC
What is the n for the combination?
DATA> 26
What is the r for the combination?
DATA> 6
The combination value is:
```

Data Display

```
--------
   230230
```

The answer is 230,230. This means that there are 230,230 possible combinations of the alphabet for items of interest with six letters. Application of combinations in calculating probability will be discussed in detail in Section 6.5 in the next chapter.

68   CHAPTER 5

The EXCEL calculation is shown below.

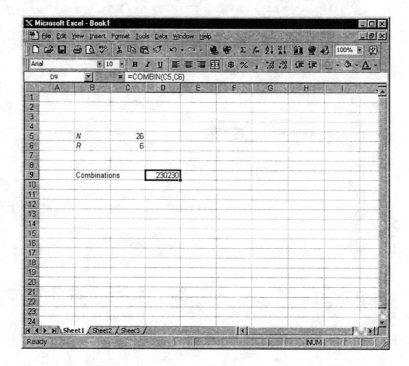

Example 5.4C

The United Jersey Bank in New Jersey is giving out gifts to depositors. If eligible, depositors may choose any 2 out of 6 gifts. How many possible combinations of gifts can different depositors select?

We will use the *comb* macro to solve this problem, given that $n = 6$ and $r = 2$.

```
MTB > %comb
Executing from file: comb.MAC
What is the n for the combination?
DATA> 6
What is the r for the combination?
DATA> 2
The combination value is:
```

Data Display

```
--------
    15
```

From the *comb* macro we can see that there are 15 possible combinations of gifts.

The EXCEL calculation is shown below.

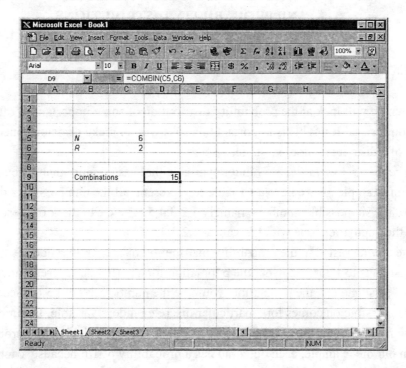

5.5 Statistical Summary

This chapter begins our study of inferential statistics. We first look at the very basic concepts of probability, because inferential statistics needs to use probability to make inferences about a population from sample data. We showed how probabilities can be calculated by using a percentage table approach. Then we discussed both permutations and combinations, which are often useful in calculating probabilities.

CHAPTER 6
DISCRETE RANDOM VARIABLES
AND PROBABILITY DISTRIBUTIONS

6.1 Introduction

In this chapter we will look at the probability distribution of some important or often used discrete random variables. This is important, because we can make decisions once we know the distribution of a discrete random variable.

A *discrete random variable* is a random variable that has no more than a countable number of values. For example, if we have a variable to represent all the numbers of a die, then the outcomes are the six possible numbers, namely 1,2,3,4,5,6. We can think of these possible numbers as the population, one with the values 1,2,3,4,5,6.

Often we are interested in the probability distribution of a discrete random variable. The first thing we must do is to figure out what the population for a particular discrete random variable is. In the case of tossing a die, the population would be the numbers 1,2,3,4,5,6. Second, we have to figure out the possible outcomes for a particular discrete random variable. In our die example it would be the individual numbers 1,2,3,4,5,6. Third, we have to calculate the probability of each outcome. In our die example, each outcome has a probability of 1/6 because each outcome has an equal chance of occurring.

Therefore, the probability of a 1 occurring from tossing a die would be 1/6. The probability of a 2 occurring would be 1/6, that of a 3 would be 1/6, that of a 4 occurring would be 1/6, that of a 5 occurring would be 1/6, that of a 6 occurring would be 1/6.

A probability distribution is a listing of all probabilities of the outcomes of a population. For our die, the probability distribution would then be

Probability Distribution of a Die	
#	probability
1	16.667%
2	16.667%
3	16.667%
4	16.667%
5	16.667%
6	16.667%
Total Prob	100% *

* number rounded

DISCRETE RANDOM VARIABLES AND PROBABILITY DISTRIBUTIONS

As indicated in the previous chapter, the sum of all the probabilities of all the outcomes of a population must equal 100%.

We can also analyze a probability distribution by graphing it. A graph of the die distribution is shown below.

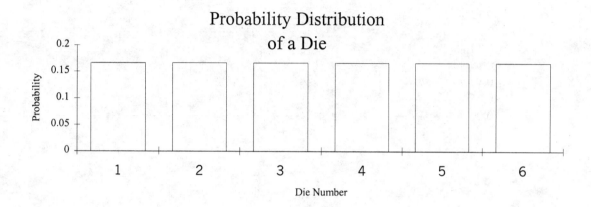

Once we know the probability distribution of a variable, we can make decisions based on it. For example, if we were going to gamble on a fair die where every number has an equal chance of occurring, we would be indifferent as to which of the six numbers to bet. But suppose we knew that the number 4 had a 50% chance of occurring and all the other numbers of the die had a 10% chance of occurring. Then we would be greatly influenced to choose the number 4 and to avoid choosing the other numbers.

The probability distribution of a die is flat, but probability distributions do not all have to look flat. Suppose instead that we have the following data on the number of defective tires that roll off a production line in a day.

Defects	Probability
0	0.05
1	0.15
2	0.20
3	0.25
4	0.25
5	0.10

The probability distribution would look like the following.

Defective Tires
Probability Distribution

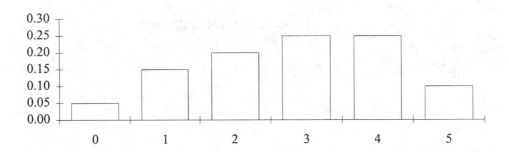

6.2 Cumulative Density Function

One question that we might ask about these tires is, "What is the chance that we might get 2 or fewer defects?'" Looking at the graph we can see that the probability of getting 0 defects is 5%, the probability of getting 1 defect is 15%, and the probability of getting 2 defects is 20%. To find the probability of getting 2 or fewer, we sum up all the probabilities of 0, 1, and 2 defects. This would be 40%. We can also ask the question, "What is the probability of getting 4 defects or fewer?'" The answer would be to add the probability of getting 3 defects to the probability of getting 2 defects or fewer and then add the probability of getting 4 defects. Mathematically it would be $0.05 + 0.15 + 0.20 + 0.25 + 0.25 = 0.9$. This process forms the concept of the *cumulative distribution function*. MINITAB can tabulate the cumulative distribution function with the *cdf* command. This command is illustrated below with the defective tire data. The first thing we should do is to put the defect data in column one of a worksheet and the probability of these defects in column two.

Below is the use of the *cdf* command to calculate the cumulative distribution function for the defective tires.

```
MTB > set c1
DATA> 0 1 2 3 4 5
DATA> end
MTB > set c2
DATA> 0.05 0.15 0.20 0.25 0.25 0.10
DATA> end
MTB > cdf;
SUBC> discrete c1 c2.
```

DISCRETE RANDOM VARIABLES AND PROBABILITY DISTRIBUTIONS

Cumulative Distribution Function

```
Discrete distribution using values in C1 and probabilities in C2

      x        P( X <= x)
   0.00          0.0500
   1.00          0.2000
   2.00          0.4000
   3.00          0.6500
   4.00          0.9000
   5.00          1.0000
```

The first column of the output shows the possible outcomes. The second column of the output shows the cumulative probability function. We can see that the probability of getting 2 defects or fewer and the probability of getting 4 defects or fewer agrees with our previous calculations.

Notice that our *cdf* command includes the *discrete* subcommand. The *discrete* subcommand tells the *cdf* command to calculate using a discrete distribution. The first argument, 'c1', is the location of the possible values of the discrete variable. The second argument, 'c2', is the location of the probability of the possible outcomes of the discrete variable.

6.3 Mean of a Discrete Variable

The mean of a discrete variable, often called the expected value, is mathematically defined as

$$\mu = \sum_{i=1}^{N} x_i P(x_i)$$

Example 6.3A

Suppose a stock analyst derives the following probability distribution for the earnings per share (EPS) of a firm.

EPS	$P(x)$
1.50	0.05
1.75	0.30
2.00	0.35
2.25	0.15
2.50	0.10
2.75	0.05

Calculate the expected value or mean using MINITAB.

To do this we would put the EPS data in column c1 and the $P(x)$ data in column c2 and store the mean in the constant k1.

```
MTB > set c1
DATA> 1.50 1.75 2.00 2.25 2.50 2.75
DATA> end
MTB > set c2
DATA> 0.05 0.30 0.35 0.15 0.10 0.05
DATA> end
MTB > let k1 = sum(c1*c2)
MTB > print k1
```

Data Display

```
K1    2.02500
```

MINITAB calculated the expected earnings per share as 2.025.

In EXCEL its possible to calculate the mean by using the *SUMPRODUCT* function. This is shown below.

Example 6.3B

Suppose the distribution of the ages of students in a class is

AGE	$P(x)$
20	0.06
21	0.10
22	0.28
23	0.39
24	0.10
25	0.03
26	0.04

Use MINITAB to calculate the mean age.

We will put the age in column c1 and probability in column c2 and the mean in the constant k1.

```
MTB > set c1
DATA> 20 21 22 23 24 25 26
DATA> end
MTB > set c2
DATA> 0.06 0.10 0.28 0.39 0.10 0.03 0.04
DATA> end
MTB > let k1 = sum(c1*c2)
MTB > print k1
```

Data Display

```
K1    22.6200
```

MINITAB calculated the mean age as 22.62.

6.4 Variance of a Discrete Variable

The formula for the variance of a discrete random variable is

$$\sigma^2 = \sum_{i=1}^{N} x_i^2 P(x_i) - \mu^2$$

Example 6.4A

Suppose the following table gives the number of defective tires that roll off a production line in a day.

DEFECTS	$P(x)$
0	0.05
1	0.15
2	0.20
3	0.25
4	0.25
5	0.10

Use MINITAB to calculate the variance.

We will name column c1 '*defects*', column c2 '*p(x)*', and column c3 '*variance*'. Below is an illustration of the MINITAB commands to calculate the variance.

```
MTB > name c1 'defects'
MTB > name c2 'p(x)'
MTB > name c3 'variance'
MTB > set c1
DATA> 0 1 2 3 4 5
DATA> end
MTB > set c2
DATA> 0.05 0.15 0.20 0.25 0.25 0.10
DATA> end
MTB > let c3(1)=sum((c1**2)*c2)-sum(c1*c2)**2
MTB > print c1-c3
```

Data Display

Row	defects	p(x)	variance
1	0	0.05	1.86
2	1	0.15	
3	2	0.20	
4	3	0.25	
5	4	0.25	
6	5	0.10	

Example 6.4B

Suppose we have the following lending rates and corresponding probabilities.

Lending,%	$P(x)$
15.00	0.100
17.00	0.075
20.00	0.075
13.00	0.200
15.00	0.150
18.00	0.150
11.00	0.100
13.00	0.075
16.00	0.075

We will put the lending rates in column c1 and the probabilities in column c2. We will name column c1 '*lending*', and column c2 '*p(x)*', and column c3 '*variance*'. The following MINITAB commands will calculate the variance for the lending rates.

```
MTB > name c1 'lending'
MTB > name c2 'p(x)'
MTB > name c3 'variance'
MTB > set c1
DATA> 15.00 17.00 20.00 13.00 15.00 18.00 11.00 13.00 16.00
DATA> end
MTB > set c2
DATA> 0.100 0.075 0.075 0.200 0.150 0.150 0.100 0.075 0.075
DATA> end
MTB > let c3(1)=sum((c1**2)*c2)-sum(c1*c2)**2
MTB > print c1-c3
```

Data Display

Row	lending	p(x)	variance
1	15	0.100	6.29
2	17	0.075	
3	20	0.075	
4	13	0.200	
5	15	0.150	
6	18	0.150	
7	11	0.100	
8	13	0.075	
9	16	0.075	

From the output we can see that the variance of the lending rates is 6.29 %.

6.5 Binomial Random Variable

We will now look at some of the important discrete random variables. One of these is called a *binomial random variable*, because the variable can only have two outcomes. Examples of binomial random variables are heads or tails in tossing a coin and defect or no defect of a product. We will call one of the outcomes an S or success, and the other outcome an F or failure. The sum of these two probabilities should equal 1. That is, $P(S) + P(F) = 1$.

Example 6.5A

Suppose we are interested in finding out the probability of getting four tails when we flip a coin five times. We will assume that the coin toss is fair, so the probability of a head is 0.5 and the probability of a tail is 0.5. For this problem, which follows a binomial probability distribution, we use the following mathematical formula:

$$\frac{n!}{r!(n-r)!}p^r q^{n-r}$$

In the formula n is how many times we flip the coin, r is how many times we want the outcome to be tails, p is the probability of success, and q is the probability of failure. Substituting all the values into the above mathematical formula would give us the following:

$$\frac{5!}{4!(5-4)!} * 0.5^4 * 0.5^{5-4} = \frac{5*4*3*2*1}{4*3*2*1*1} * 0.5^4 * 0.5^{5-4} = \frac{120}{24} * 0.03125 = 0.1562$$

This can be done quickly with MINITAB by the *pdf* command.

```
MTB > pdf;
SUBC> binomial n=5 p=.5.
```

Probability Density Function

```
Binomial with n = 5 and p = 0.500000
        x        P( X = x)
        0         0.0312
        1         0.1562
        2         0.3125
        3         0.3125
        4         0.1562
        5         0.0312
```

If we look at column x of the output for the number 4, we can see that the probability is 0.1562, which is the number that we calculated.

To get this, we used the *pdf* command and then used the subcommand *binomial* to indicate the distribution we want. The first argument indicates the size of the trial and the second argument is the probability of success.

In EXCEL we use the *binomdist* function to calculate the probability of success for a binomial distribution.

Number_s is the number of success trials
Trials is the number of independent trials
Probability_s is the number of success on each trial
Cumulative True for cumulative calculation and False for non-cumulative calculation

Example 6.5B

Assume that an insurance sales agent believes that the probability of her making a sale on a sales call is 0.20. What is the probability of success for 0,1,2,3,4, and 5 contacts? Use MINITAB to find out.

```
MTB> pdf;
SUBC> binomial n=5 p=.2.
```

Probability Density Function

Binomial with n = 5 and p = 0.200000

x	P(X = x)
0	0.3277
1	0.4096
2	0.2048
3	0.0512
4	0.0064
5	0.0003

From the MINITAB output we can see the probabilities of success for 0,1,2,3,4, and 5 sales calls.

DISCRETE RANDOM VARIABLES AND PROBABILITY DISTRIBUTIONS

Example 6.5C

A shipment of 800 calculator chips arrives at Century Electronics. The contract specifies that Century will accept this lot if a sample of 20 drawn from the shipment has no more than 1 defective chip. What is the probability of accepting the lot by applying this criterion if, in fact, 5% of the whole lot (40 chips) turns out to be defective? What if 10 % of the lot is defective?

This is a binomial situation where there are $n = 20$ trials, and $p =$ the probability of success (chip is defective) $= 0.05$. The shipment is accepted if the number of defectives is either 0 or 1. Use MINITAB to find out the probability of the shipment being accepted.

```
MTB > cdf;
SUBC> binomial 20 .05.
```

Cumulative Distribution Function

Binomial with n = 20 and p = 0.0500000

```
       x       P( X <= x)
       0         0.3585
       1         0.7358
       2         0.9245
       3         0.9841
       4         0.9974
       5         0.9997
       6         1.0000
```

From this MINITAB output we can see that the probability of acceptance when the true defect rate is 5% would be 73.58%.

From the above MINITAB output we can see that the probability of acceptance when the true defect rate is 10% would be 39.17%.

The calculation done in EXCEL using the function wizard is shown below.

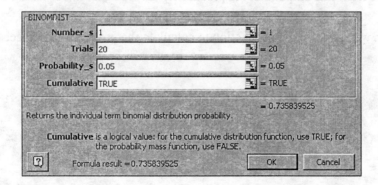

CHAPTER 6

6.6 Poisson Random Variable

The Poisson random distribution is useful for determining the probability that a particular event will occur a certain number of times over a specified time period or within the space of a particular interval. The formula for the density function of the Poisson distribution is

$$\frac{e^{-\lambda}\lambda^x}{x!}$$

Example 6.6A

Suppose a factory examines its work injury history and discovers that the chance of an accident on a given work day follows a Poisson distribution. The average number of injuries per work day is 0.01. What is the probability that there will be 3 work injuries in a given month (assume a 30-day month)?

In this problem $m = 30 * 0.01 = 0.3$.

```
MTB > pdf;
SUBC> poisson m=.3.
```

Probability Density Function
```
Poisson with mu = 0.300000

        x         P( X = x)
        0          0.7408
        1          0.2222
        2          0.0333
        3          0.0033
        4          0.0003
        5          0.0000
```

We can see from this output that the probability is .0033 that there will be 3 injuries in a given month.

In EXCEL we use the *poisson* function to calculate the probability of success for a Poisson distribution.

DISCRETE RANDOM VARIABLES AND PROBABILITY DISTRIBUTIONS

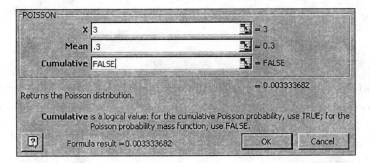

X is the number of events
Mean is the expected numeric value
Cumulative True for cumulative calculation and False for non-cumulative calculation

Example 6.6B

The average defect rate in a spark plug assembly line is 2 parts per week. Use MINITAB to figure out the probability of having more than 4 defects.

```
MTB > cdf;
SUBC> poisson 2.
```

Cumulative Distribution Function

```
Poisson with mu = 2.00000

        x       P( X <= x)
        0         0.1353
        1         0.4060
        2         0.6767
        3         0.8571
        4         0.9473
        5         0.9834
        6         0.9955
        7         0.9989
        8         0.9998
        9         1.0000


MTB > let k1=1-.9473
MTB > print k1
```

Data Display

K1   0.0527000

From the MINITAB output we can see that the cumulative probability for 4 mistakes is 0.9473. So the cumulative probability of more than 4 defects would be 1 – 0.9473 = 0.0527.

6.7 Statistical Summary

In this chapter we saw that a *discrete random variable* is a variable that has a countable number of outcomes. We also saw that a *probability distribution* is a list of all the probabilities for the outcomes of a variable. We looked at the mean and variance of a discrete variable. We also looked at two important discrete distributions, the binomial distribution and the Poisson distribution.

CHAPTER 7
THE NORMAL AND
LOGNORMAL DISTRIBUTIONS

7.1 Introduction

In the last chapter we examined the probability distribution of discrete random variables. In this chapter we will look at the probability distribution of *continuous random variables*.

Unlike the values of discrete random variables, which are limited to a finite or countable number of distinct (integer) values, values of continuous variables are *not* limited to being integers; theoretically, they are infinitely divisible. A *continuous random variable* may take on any value within an interval.

7.2 Uniform Distribution

One interesting characteristic of continuous random variables is that the entire range of probabilities under a continuous random variable has a total area equal to one. This is most clearly shown in the uniform continuous distribution. The formula for the density function of the uniform distribution is

$$f(x) = \frac{1}{d-c}, c < x < d$$

The c is the beginning point and the d is the ending point of the uniform distribution. A graph of the uniform distribution is shown below.

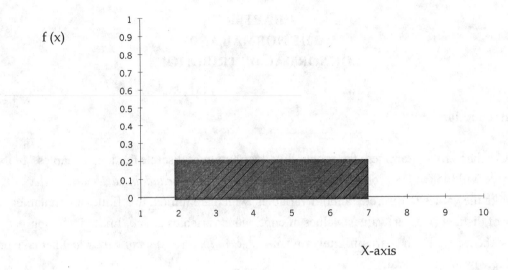

This is a uniform distribution because any two equal intervals of the same size will have an equal probability. This uniform probability density function is also supposed to have a total area equal to one. We can confirm this by using geometry. The area of a rectangle is height times length; so in this case the area is 0.2 * 5 = 1.

We can also create the continuous uniform distribution using MINITAB.

```
MTB > Set c1
DATA>    1( 1 : 10 / .01 )1
DATA>    End.
MTB > pdf c1 c2;
SUBC> uniform 2 7.
MTB > Plot c2*c1;
SUBC>    Connect;
SUBC>    Title "Uniform Probability Density Function";
SUBC>    Axis 1;
SUBC>       Label "X-Axis";
SUBC>    Axis 2;
SUBC>       Label "f(x)".
```

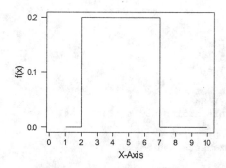

We can confirm that this plot has an area that is equal to 1 by seeing that the height is 0.2 and the width is 5. If we multiply these two numbers together, we get 1.

We can also see that the area is equal to 1 by calculating the uniform cumulative distribution function. Doing this shows us that at point 7 the uniform cumulative distribution function equals 1.

```
MTB > cdf c1 c3;
SUBC> uniform 2 7.
MTB > Plot c3*c1;
SUBC>    Connect;
SUBC>    Title "Uniform Cumulative Density Function";
SUBC>    Axis 1;
SUBC>      Label "x-axis";
SUBC>    Axis 2;
SUBC>      Label "f(x)".
```

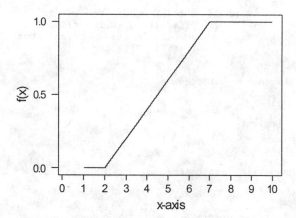

Below is the programming code in EXCEL to create a Uniform graph. The code is based on the formula for the density function of the uniform distribution.

CHAPTER 7

```vb
'/**********************************************************
'/*Purpose: Create a Uniform distribution  with intervals
'/          2 and 7
'/**********************************************************
Sub UniformPDF()
    'Define Variables
    Dim Ynumbers() As Variant
    Dim Xnumbers() As Variant
    Dim Yobservations As Integer
    Dim nCounter As Integer
    Dim chartUniformPDF As Object
    Dim wsDataForChart As Worksheet
    Dim objNS As Object ' store adding new series

    'add workhsheet to hold chartdata
    Set wsDataForChart = Worksheets.Add

    'add label for data
    wsDataForChart.Range("a1") = "X-Values"
    wsDataForChart.Range("b1") = "Uniform PDF"

    '9 intervals with 100 .01 per interval
    Yoberservations = (10 - 1) * 100

    'reset the size of the Y and X arrays
    ReDim Ynumbers(1 To Yoberservations)
    ReDim Xnumbers(1 To Yoberservations)

    'calculate the uniform distribution
    For nCounter = 1 To Yoberservations
        wsDataForChart.Range("a1").Offset(nCounter, 0).Value = 0.99 + 0.01 * nCounter
        If (0.99 + 0.01 * nCounter) >= 2 And (0.99 + 0.01 * nCounter) <= 7 Then
            'set Uniform distribution values
            wsDataForChart.Range("b1").Offset(nCounter, 0).Value = 1 / (7 - 2)
        Else
            'outside of interval so value is zero
            wsDataForChart.Range("b1").Offset(nCounter, 0).Value = 0
        End If
    Next

    'set a variable pointing to the new chart
    Set chartUniformPDF = Charts.Add
    With chartUniformPDF
        .ChartType = xlLine
        .SeriesCollection(1).XValues = wsDataForChart.Range("a1").CurrentRegion. _
                            Offset(1, 0).Columns(1)
        .SeriesCollection(1).Values = wsDataForChart.Range("a1").CurrentRegion. _
                            Offset(1, 0).Columns(2)
        .HasLegend = False
        With .PlotArea
            .Interior.ColorIndex = xlNone
            .ClearFormats
        End With
        With .Axes(xlCategory)
            .HasTitle = False
            .TickLabelSpacing = 100
            .TickMarkSpacing = 100
        End With
        .Axes(xlValue).MajorGridlines.Delete
        'set chart titles
        .HasTitle = True
        With .ChartTitle
            .Characters.Text = "Uniform Probability Density Function"
            .Font.Size = 26
        End With
    End With
End Sub
```

Running the above EXCEL programming code creates the graph shown below.

Uniform Probability Density Function

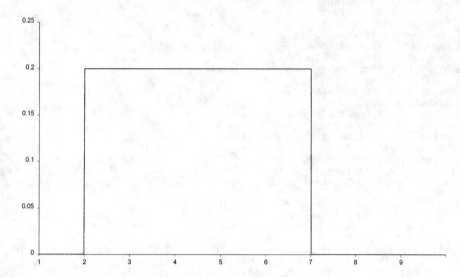

We will now create the cumulative uniform graph in EXCEL. To create this graph we based it on the cumulative uniform density function below.

$$F(X \leq x) \begin{cases} 0 & x < c \\ \dfrac{x-c}{d-c} & c \leq x \leq d \\ 1 & x > d \end{cases}$$

The programming code to graph the cumulative uniform density function in EXCEL is shown below.

```vb
'/**********************************************************
'/*Purpose: Create a cumulative Uniform distribution
'/          with intervals 2 and 7
'/**********************************************************
Sub CulmUniformPDF()
    'Define Variables
    Dim Ynumbers() As Variant
    Dim Xnumbers() As Variant
    Dim Yobservations As Integer
    Dim nCounter As Integer
    Dim chartUniformPDF As Object
    Dim wsDataForChart As Worksheet
    Dim objNS As Object ' store adding new series
    
    'add workhsheet to hold chartdata
    'add label for data
    wsDataForChart.Range("a1") = "X-Values"
    wsDataForChart.Range("b1") = "Uniform PDF"
    '9 intervals with 100 .01 per interval
    Yoberservations = (10 - 1) * 100
    'reset the size of the Y and X arrays
    ReDim Ynumbers(1 To Yoberservations)
    ReDim Xnumbers(1 To Yoberservations)
    
    'calculate the culmulative uniform distribution
    For nCounter = 1 To Yoberservations
        wsDataForChart.Range("a1").Offset(nCounter, 0).Value = 0.99 + 0.01 * nCounter
        'set Cumulative Uniform distribution values
        If (0.99 + 0.01 * nCounter) >= 2 And (0.99 + 0.01 * nCounter) <= 7 Then
            wsDataForChart.Range("b1").Offset(nCounter, 0).Value = _
                ((0.99 + 0.01 * nCounter) - 2) / (7 - 2)
        ElseIf (0.99 + 0.01 * nCounter) < 2 Then
            'outside of interval so value is zero
            wsDataForChart.Range("b1").Offset(nCounter, 0).Value = 0
        ElseIf (0.99 + 0.01 * nCounter) > 7 Then
            wsDataForChart.Range("b1").Offset(nCounter, 0).Value = 1
        End If
    Next
    
    'set a variable pointing to the new chart
    Set chartUniformPDF = Charts.Add
    With chartUniformPDF
        .ChartType = xlLine
        .SeriesCollection(1).XValues = wsDataForChart.Range("a1").CurrentRegion. _
                            Offset(1, 0).Columns(1)
        .SeriesCollection(1).Values = wsDataForChart.Range("a1").CurrentRegion. _
                            Offset(1, 0).Columns(2)
        
        .HasLegend = False
        With .PlotArea
            .Interior.ColorIndex = xlNone
            .ClearFormats
        End With
        With .Axes(xlCategory)
            .HasTitle = False
            .TickLabelSpacing = 100
            .TickMarkSpacing = 100
        End With
        .Axes(xlValue).MajorGridlines.Delete
        'set chart titles
        .HasTitle = True
        With .ChartTitle
            .Characters.Text = "Uniform Cumulative Density Function"
            .Font.Size = 26
        End With
    End With
End Sub
```

Running the above EXCEL programming code creates the following graph,

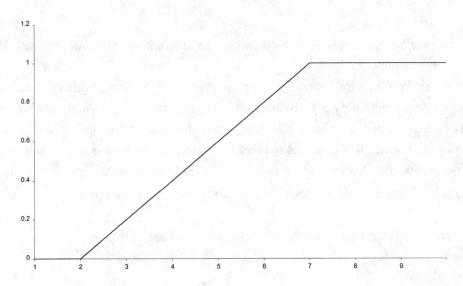

7.3 Normal Distribution

Continuous variables have many shapes, but one of the most important is the normal random variable. This density function of the normal distribution looks like the following.

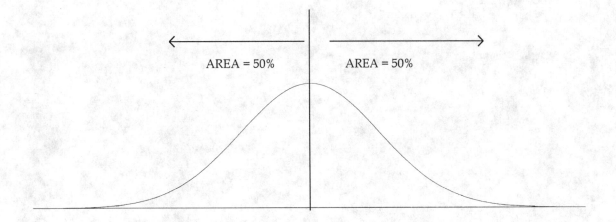

The formula for this function is

$$f(x) = \frac{1}{\sqrt{2\pi}\sigma} e^{-(x-\mu)^2/2\sigma^2}, \quad -\infty < x < \infty$$

We can see that this could be very complicated and tedious to calculate, but with MINITAB it is no longer a problem.

We should keep in mind some of the characteristics of the normal distribution. First, the distribution goes indefinitely in both directions. The area under the normal distribution is equal to 1. The normal distribution has a bell shape where the greatest probability density is at the midpoint. The normal distribution is symmetrical at its midpoint. That is, the probability to the left of the normal distribution is equal to 0.5 and the probability to the right is also equal to 0.5. The shape of the normal distribution is influenced by the standard deviation, σ, and the mean, μ.

The MINITAB commands for the normal distribution are as follows:

```
MTB > Set c1
DATA>   1( -10 : 40 / 1 )1
DATA>   End.
MTB > PDF c1 c2;
SUBC>   Normal 15 5.
MTB > Plot C2*c1;
SUBC>   Connect;
SUBC>   Title "Normal Probability Density Function";
SUBC>   Axis 1;
SUBC>     Label "x-axis";
SUBC>   Axis 2;
SUBC>     Label "y-axis".
```

Normal Probability Density Function

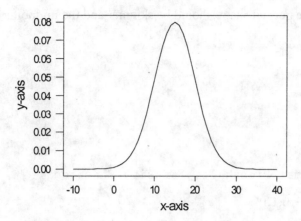

We will now calculate and plot the cumulative density function of the normal distribution to demonstrate that the area or total probability of a normal distribution is equal to 1.

```
MTB > cdf c1 c3;
SUBC> normal 15 5.
MTB > Plot C3*c1;
SUBC>    Connect;
SUBC>    Title 'Cumulative Normal Distribution Function';
SUBC>    Axis 1;
SUBC>      Label "x-axis".
```

Cumulative Normal Distribution Function

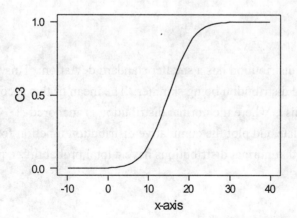

To illustrate that the normal distribution is influenced by σ and μ, we will create another normal distribution with different σ and μ. This time we will have a smaller σ and a larger μ. This would mean that the normal shape should be skinnier and the midpoint would be at the larger μ.

```
MTB > Set c1
DATA>    1( 25 : 45 / .01 )1
DATA>    End.
MTB > pdf c1 c2;
SUBC> normal 35 2.
MTB > Plot c2*c1;
SUBC>    Connect;
SUBC>    Title "Normal Probability Density Function";
SUBC>    Axis 1;
SUBC>      Label "x-axis";
SUBC>    Axis 2;
SUBC>      Label "f(x)".
```

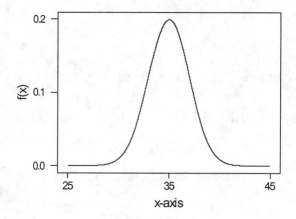

This second normal distribution has a smaller standard deviation. This is illustrated in the above plot by the normal distribution being skinnier. The mean of this second normal distribution is 35, and this is where the normal distribution is anchored.

We will again calculate and plot the cumulative distribution function for the second normal distribution to illustrate that normal distributions have a total probability equal to 1.

```
MTB > cdf c1 c3;
SUBC> normal 35 2.
MTB > Plot c3*c1;
SUBC>   Connect;
SUBC>   Title "Cumulative Distribution Function";
SUBC>   Axis 1;
SUBC>     Label "x-axis";
SUBC>   Axis 2;
SUBC>     Label "f(x)".
```

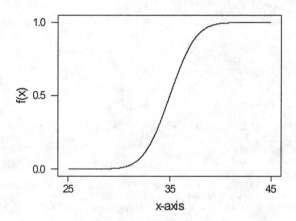

Below is the programming code in EXCEL to create a Normal graph. This programming code will product a normal distribution with a mean of 15 and a standard deviation of 5.

CHAPTER 7

```vba
'/***********************************************************
'/*Purpose: Create a Normal Probablity Density Function
'/         with mean 15 and standard deviation 5
'/***********************************************************
Sub NormalPDF()
    'Define Variables
    Dim Ynumbers() As Variant
    Dim Xnumbers() As Variant
    Dim Yobservations As Integer
    Dim nCounter As Integer
    Dim chartUniformPDF As Object
    Dim wsDataForChart As Worksheet
    Dim objNS As Object ' store adding new series
    
    'add workhsheet to hold chartdata
    Set wsDataForChart = Worksheets.Add
    
    'add label for data
    wsDataForChart.Range("a1") = "X-Values"
    wsDataForChart.Range("b1") = "Uniform PDF"
    
    Yoberservations = (40 - (-10)) + 1
    
    'reset the size of the Y and X arrays
    ReDim Ynumbers(1 To Yoberservations)
    ReDim Xnumbers(1 To Yoberservations)
    
    'calculate the normal distribution
    For nCounter = -10 To 40
        wsDataForChart.Range("a1").Offset(nCounter + 11, 0).Value = nCounter
        'set  normal distribution values
        'first argument is x value
        'second argument is mean
        'third argument is variance
        'fourth argument is false to calculate normal pdf and true to calc cumulative
        wsDataForChart.Range("b1").Offset(nCounter + 11, 0).Value = _
            WorksheetFunction.NormDist(wsDataForChart.Range("a1"). _
Offset(nCounter + 11, 0).Value, 15, 5, False)
    Next
    
    'set a variable pointing to the new chart
    Set chartUniformPDF = Charts.Add
    With chartUniformPDF
        .ChartType = xlLine
        .SeriesCollection(1).XValues = wsDataForChart.Range("a1").CurrentRegion. _
                                Offset(1, 0).Columns(1)
        .SeriesCollection(1).Values = wsDataForChart.Range("a1").CurrentRegion. _
                                Offset(1, 0).Columns(2)
        .HasLegend = False
        With .PlotArea
            .Interior.ColorIndex = xlNone
            .ClearFormats
        End With
        With .Axes(xlCategory)
            .HasTitle = False
            .TickLabelSpacing = 2
            .TickMarkSpacing = 2
        End With
        .Axes(xlValue).MajorGridlines.Delete
        'set chart titles
        .HasTitle = True
        With .ChartTitle
            .Characters.Text = "Normal Probability Density Function"
            .Font.Size = 26
        End With
    End With
End Sub
```

Running the above code produces the following graph:

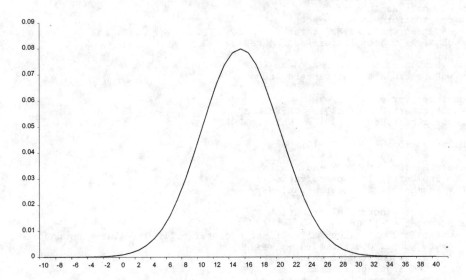

Normal Probability Density Function

The following is the EXCEL program to create a cumulative normal distribution function.

CHAPTER 7

```vba
'/**********************************************************
'/*Purpose: Create a Cumulative Normal Distribution Function
'/         with mean 15 and standard deviation 5
'/**********************************************************
Sub CumNormal()
    'Define Variables
    Dim Ynumbers() As Variant
    Dim Xnumbers() As Variant
    Dim Yobservations As Integer
    Dim nCounter As Integer
    Dim chartUniformPDF As Object
    Dim wsDataForChart As Worksheet
    Dim objNS As Object ' store adding new series

    'add workhsheet to hold chartdata
    Set wsDataForChart = Worksheets.Add

    'add label for data
    wsDataForChart.Range("a1") = "X-Values"
    wsDataForChart.Range("b1") = "Uniform PDF"

    Yoberservations = (40 - (-10)) + 1

    'reset the size of the Y and X arrays
    ReDim Ynumbers(1 To Yoberservations)
    ReDim Xnumbers(1 To Yoberservations)

    'calculate the normal distribution
    For nCounter = -10 To 40
        wsDataForChart.Range("a1").Offset(nCounter + 11, 0).Value = nCounter
        'set  normal distribution values
        'first argument is x value
        'second argument is mean
        'third argument is variance
        'fourth argument is false to calculate normal pdf and true to calc cumulative
        wsDataForChart.Range("b1").Offset(nCounter + 11, 0).Value = _
            WorksheetFunction.NormDist(wsDataForChart.Range("a1"). _
Offset(nCounter + 11, 0).Value, 15, 5, True)
    Next

    'set a variable pointing to the new chart
    Set chartUniformPDF = Charts.Add
    With chartUniformPDF
        .ChartType = xlLine
        .SeriesCollection(1).XValues = wsDataForChart.Range("a1").CurrentRegion. _
                                   Offset(1, 0).Columns(1)
        .SeriesCollection(1).Values = wsDataForChart.Range("a1").CurrentRegion. _
                                   Offset(1, 0).Columns(2)
        .HasLegend = False
        With .PlotArea
            .Interior.ColorIndex = xlNone
            .ClearFormats
        End With
        With .Axes(xlCategory)
            .HasTitle = False
            .TickLabelSpacing = 2
            .TickMarkSpacing = 2
        End With
        .Axes(xlValue).MajorGridlines.Delete
        'set chart titles
        .HasTitle = True
        With .ChartTitle
            .Characters.Text = "Cumulative Normal Distribution Function"
            .Font.Size = 26
        End With
    End With
End Sub
```

Running the above code produces the following graph:

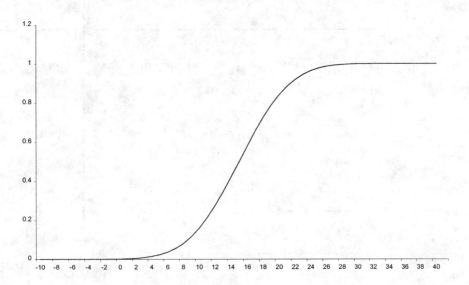

Cumulative Normal Distribution Function

Example 7.3A

Once we know that a variable follows a normal distribution, we can make a decision based on this knowledge.

A Dunkin' Donuts shop located in New Brunswick, New Jersey, sells dozens of fresh donuts. Any donuts remaining unsold at the end of the day are either discarded or sold elsewhere at a loss. The demand for the Dunkin' Donuts at this shop has followed a normal distribution with $\mu = 50$ and $\sigma = 5$ dozen. Use MINITAB to figure out how many dozen donuts this Dunkin' Donuts shop should make each day so that it can meet the demand 95 % of the time.

```
MTB > invcdf .95;
SUBC> normal 50 5.
```

Inverse Cumulative Distribution Function

```
Normal with mean = 50.0000 and standard deviation = 5.00000

 P( X <= x)         x
    0.9500       58.2243
```

From the above output we can see that the Dunkin' Donuts shop should make 58.22 dozen to meet the demand 95% of the time.

This is illustrated graphically below.

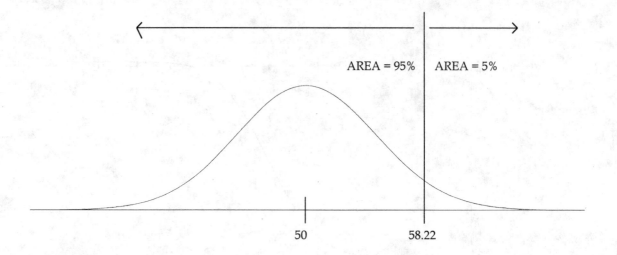

7.4 Lognormal Distribution

The shape of the density function of the lognormal distribution looks like the following:

Lognormal Distribution

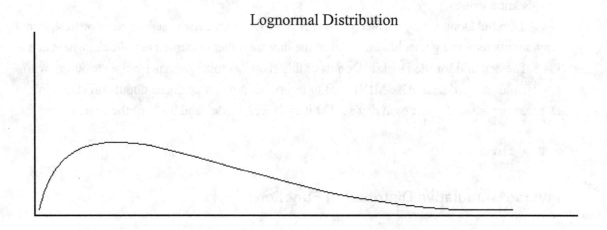

The mathematical formula for this density function is

$$f(x) = \frac{1}{x\sigma\sqrt{2\pi}} \exp[-\frac{1}{2\sigma^2}(\ln x - \mu)^2], \quad x > 0$$

As you can see, the lognormal distribution is skewed positively to the right. As with all distributions, the area under the density function is equal to 1. Below, we use MINITAB to create the lognormal distribution.

In the *lognormal* subcommand below, the first argument is the mean of the lognormal distribution and the second argument is the standard deviation of the lognormal distribution.

```
MTB > Set c1
DATA>   1( .006 : 17 / .01 )1
DATA>   End.
MTB > pdf c1 c2;
SUBC> lognormal 1 1.
MTB > Plot c2*c1;
SUBC>    Connect;
SUBC>    Title "Lognormal Density Function";
SUBC>    Axis 1;
SUBC>      Label "x-axis";
SUBC>    Axis 2;
SUBC>      Label "f(x)".
```

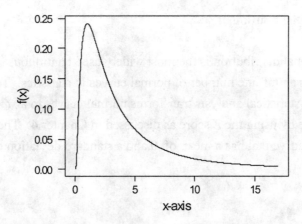

Now let us calculate and plot the lognormal cumulative distribution function to illustrate that the total area under the density function of the lognormal distribution is equal to 1.

```
MTB >cdf c1 c3;
SUBC> lognormal 1 1.
MTB > Plot c3*c1;
SUBC>   Connect;
SUBC>   Title "Cumulative Lognormal Distribution";
SUBC>   Axis 1;
SUBC>     Label "x-axis";
SUBC>   Axis 2;
SUBC>     Label "f(x)".
```

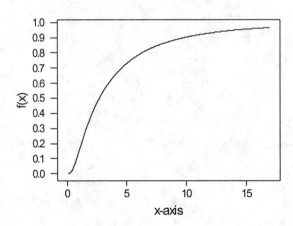

7.5 Standard Normal Distribution

We will find that the normal distribution is the most widely used continuous distribution in statistics. In fact, there is an infinite number of normal curves in statistics. To analyze normal curves more easily, most statistical analysis transforms normal curves to a *standard normal distribution*. This is done by using the Z score as discussed in Chapter 4. The *standard normal distribution* is a normal curve that has a mean of 0 and a standard deviation of 1.

7.6 Normal Approximating the Binomial

We can approximate a binomial distribution using a normal distribution if n is sufficiently large. A rule of thumb would be n equal to or greater than 30.

The mean of a binomial distribution is np. The variance of a binomial distribution is $np(1-p)$.

The fact that the normal distribution can approximate a binomial is important because we will find out that the binomial is one of the most difficult distributions to calculate when n becomes large, as illustrated in the next example. The binomial calculation in the next problem will be so big that even MINITAB cannot calculate it.

Example 7.6A

Suppose a very bumpy conveyor belt in a brewery transports beer bottles from the point where they are capped to the point where they are boxed for shipping. There is a 16 % chance that each beer bottle will fall off the conveyor belt. In 1 hour, exactly 1,000 beer bottles travel from one end of the belt to the other. Use MINITAB to find the probability that 185 beer bottles or fewer will fall off the conveyor belt.

The mean of this problem is

$$1,000 * 0.16 = 160$$

The variance of this problem is

$$1,000 * 0.16 * (1 - 0.16) = 134.4$$

MINITAB's efforts to perform the normal and binomial calculations are illustrated below.

```
MTB > let k1=sqrt(134.4)
MTB > cdf 185;
SUBC> normal 160 k1.
```

Cumulative Distribution Function

```
Normal with mean = 160.000 and standard deviation = 11.5931

         x        P( X <= x)
   185.0000        0.9845
```

```
MTB > cdf 185;
SUBC> binomial 1000 .16.

Cumulative Distribution Function

Binomial with n = 1000 and p = 0.160000

         x     P( X <= x)

* ERROR * Completion of computation impossible.
```

From the MINITAB output, the normal calculation obtains the probability of getting 185 or less as 0.9845.

However, MINITAB could not do the binomial calculation when the size of *n* was equal to 1000. The reason is the *n!* in the binomial formula. One of the calculations that MINITAB would have to make would be the factorial of 185, which has a number of more than 18 integers! Unless specifically programmed to handle this large a number, a computer program cannot calculate the binomial distribution.

In EXCEL, I used the binomial formula to calculate the probability that 185 bottles or fewer will fall off is 0.984801.

The reason that we can use a normal distribution to approximate the binomial distribution is because our sample size is very large and because of the central limit theorem, which will be discussed in Chapter 8.

This example shows why it is valuable to know that the normal distribution can approximate a binomial distribution. Without knowing this fact, we cannot solve problems where the binomial *n* is large.

Below is a graphical illustration of the above problem.

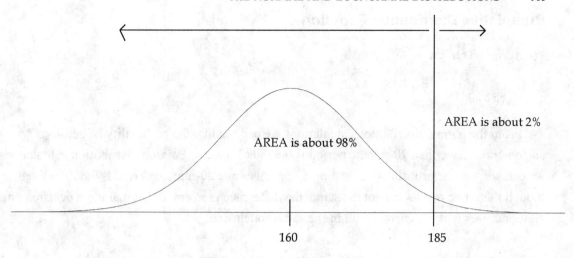

7.7 Normal Approximating the Poisson

The mean and variance of a Poisson distribution are both λ.

Example 7.7A

Assume that during a 20-minute period a bank has an average of 50 customers entering it. Suppose that we would like to find the probability of 57 customers or fewer coming into the bank within a 20-minute period. Assume that this is a Poisson random variable situation.

Our mean and variance would be 50.

```
MTB > let k1=sqrt(50)
MTB > cdf 57;
SUBC> normal 50 k1.
```

Cumulative Distribution Function

```
Normal with mean = 50.0000 and standard deviation = 7.07107

        x       P( X <= x)
   57.0000        0.8389


MTB > cdf 57;
SUBC> poisson 50.
```

Cumulative Distribution Function

```
Poisson with mu = 50.0000
       x         P( X <= x)
    57.00          0.8551
```

From the normal distribution calculations we can see that the probability of getting 57 customers or fewer in a 20-minute period is 0.8389. From the Poisson distribution calculations we can see that the probability of getting 57 or fewer in a 20-minute period is 0.8551. Even though these two results are not the same, the difference is so small that making a decision on either number will in general result in the same conclusion.

The graph of the probability of 57 customers or fewer in a 20-minute period is as follows:

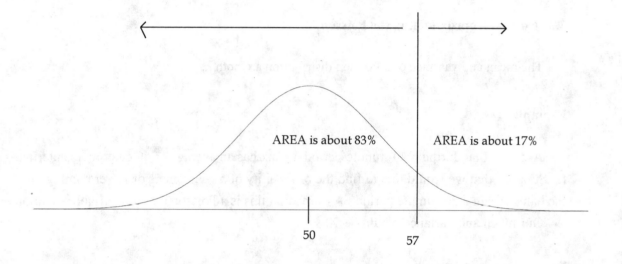

7.8 Statistical Summary

In this chapter we looked at the uniform, normal, and lognormal continuous distributions. We saw that the normal distribution is influenced by the mean and standard deviation. We also saw that the normal distribution can approximate a binomial distribution. This is valuable because when the *n* of a binomial distribution is very large, the binomial distribution becomes very hard to calculate. We also saw that the normal distribution can approximate a Poisson distribution.

CHAPTER 8
SAMPLING AND SAMPLING DISTRIBUTIONS

8.1 Introduction

In Chapter 6 we looked at the binomial distribution and the Poisson distribution. In Chapter 7 we looked at the uniform, normal, and lognormal distributions. In Chapter 9 we will look at the chi-square, F, exponential, and t distributions.

In all these distributions, one of the things that we will be interested in is the mean. One of the most important questions we can ask about the sample mean is, "What is the distribution of the sample mean?" If we take a sample from a lognormal distribution, is the distribution of the sample mean a lognormal distribution? Or, if we take a sample from an F distribution, is the distribution of the sample mean an F distribution? The answer surprisingly is no. It turns out that if the sample size is sufficiently large, the sampling distribution of a mean will approximate a normal distribution. A general rule of thumb for "sufficiently large" is a sample size greater than 30. This concept, known as the *central limit theorem,* is stated below.

> As the sample size (n) from a given population gets 'large enough,' the sampling distribution of the mean, $\bar{x}$, can be approximated by a normal distribution with mean, μ, and standard deviation, $\sigma/\sqrt{n}$, regardless of the distribution of the individual values in the population.

In this chapter we will illustrate this theorem with the distributions we considered in Chapters 6 and 7. We will also illustrate this theorem in Chapter 9 with the distributions considered there. In every case, we will see that as n increases, the distribution of the sample mean will approach a normal distribution.

MINITAB provides enormous help in illustrating the *central limit theorem.* Let us see how many calculations MINITAB will be making in this chapter and in Chapter 9.

	Number of MINITAB Calculations				
	$N = 5$	$N = 10$	$N = 30$	$N = 50$	
Distribution	300 Samples	300 Samples	300 Samples	300 Samples	TOTAL
Uniform	1,500	3,000	9,000	15,000	28,500
Normal	1,500	3,000	9,000	15,000	28,500
Exponential	1,500	3,000	9,000	15,000	28,500
Lognormal	1,500	3,000	9,000	15,000	28,500
F	1,500	3,000	9,000	15,000	28,500
Chi-square	1,500	3,000	9,000	15,000	28,500
t	1,500	3,000	9,000	15,000	28,500
Binomial	1,500	3,000	9,000	15,000	28,500
Poisson	1,500	3,000	9,000	15,000	28,500
	13,500	27,000	81,000	135,000	256,500

From the above table we can see that MINITAB is doing 256,500 calculations! Just think how much time we are saving by using MINITAB instead of doing these 256,500 calculations by hand.

For each distribution we will randomly select
1. 300 samples of size 5
2. 300 samples of size 10
3. 300 samples of size 30
4. 300 samples of size 50

From this exercise, we will see that for every distribution shown in this chapter, the sample mean distribution approximates a normal distribution. Some sample mean, distribution will have an approximate normal distribution with a sample size as small as 5, whereas other sample mean distributions will approach a normal distribution as the sample size increases.

We will use the *random* command to generate random numbers from any distribution given by its subcommand. It should be noted that since these numbers are randomly generated, an exact reproduction of the dotplot is not possible, but the general behavior of each dotplot shown below should result.

8.2 Uniform Distribution

Sample Size = 5, 300 Samples

```
MTB > random 300 c1-c5;
SUBC> uniform 5 6.
MTB > rmean c1-c5 c10
MTB > Histogram c10;
SUBC>   MidPoint;
SUBC>   NInterval 100;
SUBC>   Bar;
SUBC>   Axis 1;
SUBC>     Label "x-axis"
SUBC>   Axis 2;
SUBC>   Tick 1;
SUBC>   Tick 2.
```

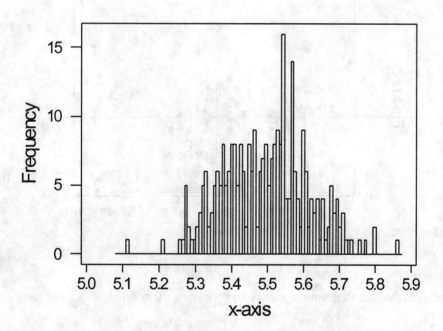

Sample Size = 10, 300 Samples

```
MTB >random 300 c1-c10;
SUBC> uniform 5 6.
MTB > rmean c1-c10 c15;
MTB > Histogram c15;
MTB >Histogram c15;
SUBC>    MidPoint;
SUBC>    NInterval 100;
SUBC>    Bar;
SUBC>    Minimum 2 0;
SUBC>    Maximum 2 15;
SUBC>    Axis 1;
SUBC>      Label "x-axis";
SUBC>    Axis 2;
SUBC>    Tick 1;
SUBC>    Tick 2.
```

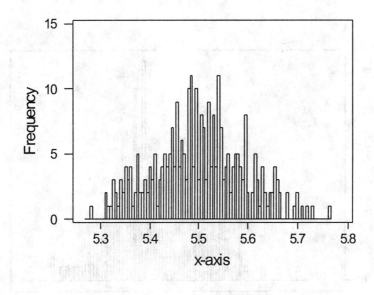

We represent the fact that sample size is 10 by having 10 columns.

Sample Size = 30, 300 Samples

```
MTB > random 300 c1-c30;
SUBC> uniform 5 6.
MTB > rmean c1-c30 c35
MTB > Histogram c35;
SUBC>   MidPoint;
SUBC>   NInterval 100;
SUBC>   Bar;
SUBC>   Minimum 2 0;
SUBC>   Maximum 2 15;
SUBC>   Axis 1;
SUBC>     Label "x-axis";
SUBC>   Axis 2;
SUBC>   Tick 1;
SUBC>   Tick 2.
```

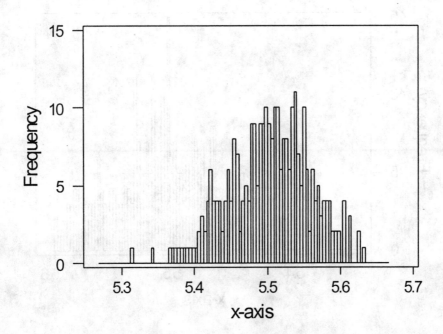

We represent the fact that sample size is 30 by having 30 columns.

Sample Size = 50, 300 Samples

```
SUBC> uniform 5 6.
MTB > rmean c1-c50 c55
MTB > Histogram c55;
SUBC>    MidPoint;
SUBC>    NInterval 100;
SUBC>    Bar;
SUBC>    Minimum 2 0;
SUBC>    Maximum 2 15;
SUBC>    Axis 1;
SUBC>      Label "x-axis";
SUBC>    Axis 2;
SUBC>    Tick 1;
SUBC>    Tick 2.
```

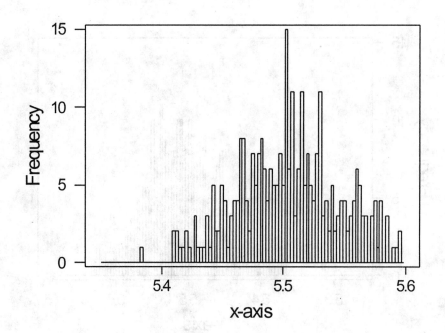

We represent the fact that sample size is 50 by having 50 columns.

8.3 Normal Distribution

In the commands issued below, the *normal* subcommand tells the *random* command to generate random numbers from the normal distribution. The first argument of the normal distribution is the mean of the distribution. The second argument is the standard deviation of the distribution.

Sample Size = 5, 300 Samples

```
MTB > rmean c1-c5 c10
MTB > Histogram C10;
SUBC>   MidPoint;
SUBC>   NInterval 100;
SUBC>   Bar;
SUBC>   Axis 1;
SUBC>     Label "x-axis";
SUBC>   Axis 2.
```

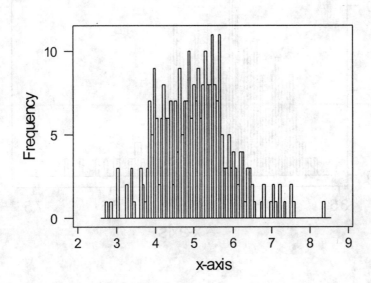

Sample Size = 10, 300 Samples

```
MTB > random 300 c1-c10;
SUBC> normal 5 2.
MTB > rmean c1-c10 c15
MTB > Histogram c15;
SUBC>    MidPoint;
SUBC>    NInterval 100;
SUBC>    Bar;
SUBC>    Axis 1;
SUBC>       Label "x-axis";
SUBC>    Axis 2.
```

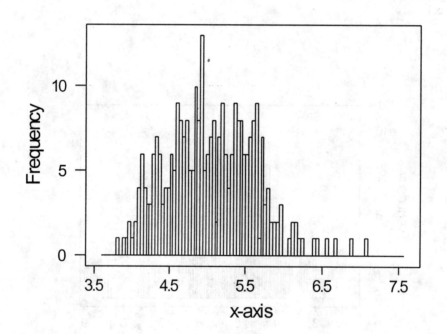

Sample Size = 30, 300 Samples

```
MTB > random 300 c1-c30;
SUBC> normal 5 2.
MTB > rmean c1-c30 c35
MTB > Histogram c35;
SUBC>   MidPoint;
SUBC>   NInterval 100;
SUBC>   Bar;
SUBC>   Axis 1;
SUBC>     Label "x-axis";
SUBC>   Axis 2.
```

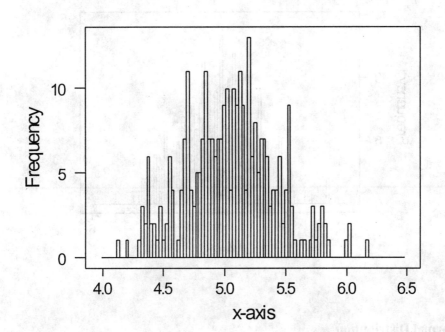

Sample Size = 50, 300 Samples

```
MTB > random 300 c1-c50;
SUBC> normal 5 2.
MTB > rmean c1-c50 c55
MTB > Histogram c55;
SUBC>   MidPoint;
SUBC>   NInterval 100;
SUBC>   Bar;
SUBC>   Axis 1;
SUBC>     Label "x-axis";
SUBC>   Axis 2.
```

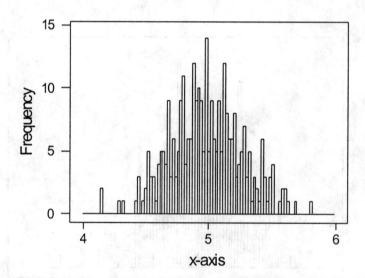

8.4 Lognormal Distribution

In the commands issued below, the *lognormal* subcommand tells the *random* command to generate numbers randomly from the lognormal distribution. The first argument of the *lognormal* subcommand is the mean. The second argument is the standard deviation.

Sample Size = 5, 300 Samples

```
MTB > random 300 c1-c5;
SUBC> lognormal 1 1.
MTB > rmean c1-c5 c10
MTB > Histogram C10;
SUBC>   MidPoint;
SUBC>   NInterval 100;
SUBC>   Bar;
SUBC>   Axis 1;
SUBC>     Label "x-axis";
SUBC>   Axis 2.
```

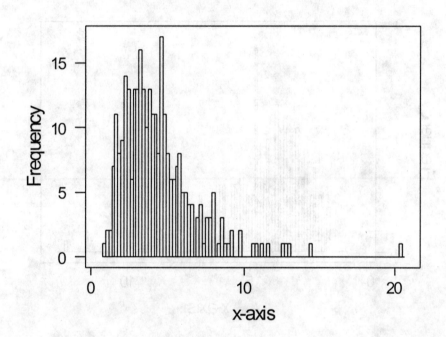

Sample Size = 10, 300 Samples

```
MTB > random 300 c1-c10;
SUBC> lognormal 1 1.
MTB > rmean c1-c10 c15
MTB > Histogram C15;
SUBC>    MidPoint;
SUBC>    NInterval 100;
SUBC>    Bar;
SUBC>    Axis 1;
SUBC>       Label "x-axis";
SUBC>    Axis 2.
```

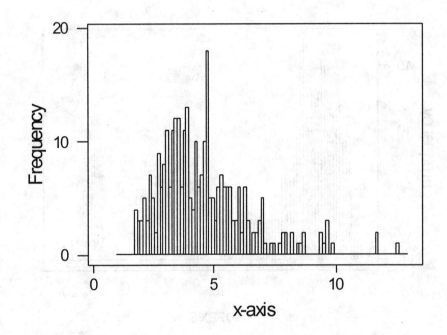

Sample Size = 30, 300 Samples

```
MTB > random 300 c1-c30;
SUBC> lognormal 1 1.
MTB > rmean c1-c30 c35
MTB > Histogram C35;
SUBC>   MidPoint;
SUBC>   NInterval 100;
SUBC>   Bar;
SUBC>   Axis 1;
SUBC>     Label "x-axis";
SUBC>   Axis 2.
```

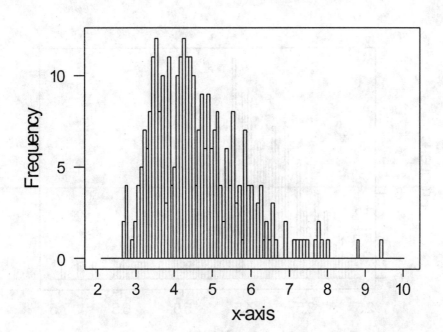

Sample Size = 50, 300 Samples

```
MTB >   random 300 c1-c50;
SUBC> lognormal 1 1.
MTB > rmean c1-c50 c55
MTB > Histogram c55;
SUBC>    MidPoint;
SUBC>    NInterval 100;
SUBC>    Bar;
SUBC>    Axis 1;
SUBC>      Label "x-axis";
SUBC>    Axis 2.
```

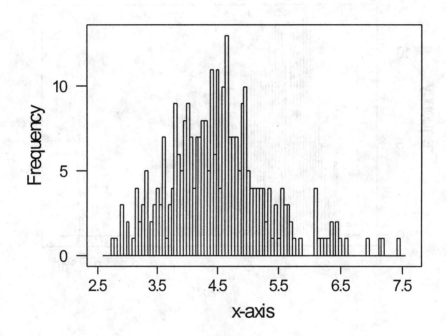

8.5 Binomial Distribution

In the commands issued below, the *binomial* subcommand tells the *random* command to generate random numbers from the binomial distribution. The first argument of the *binomial* subcommand is the number of trials. The second argument is the probability.

Sample Size = 10, 300 Samples

```
MTB > random 300 c1-c5;
SUBC> binomial n=100 p=.3.
MTB > rmean c1-c5 c10
MTB > Histogram c10;
SUBC>   MidPoint;
SUBC>   NInterval 100;
SUBC>   Bar;
SUBC>   Axis 1;
SUBC>     Label "x-axis";
SUBC>   Axis 2.
```

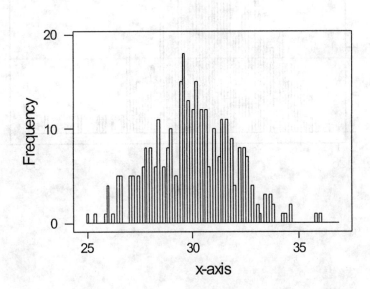

Sample Size = 10, 300 Samples

```
MTB > random 300 c1-c10;
SUBC> binomial n=100 p=.3.
MTB > rmean c1-c10 c15
MTB > Histogram c15;
SUBC>   MidPoint;
SUBC>   NInterval 100;
SUBC>   Bar;
SUBC>   Axis 1;
SUBC>     Label "x-axis";
SUBC>   Axis 2.
```

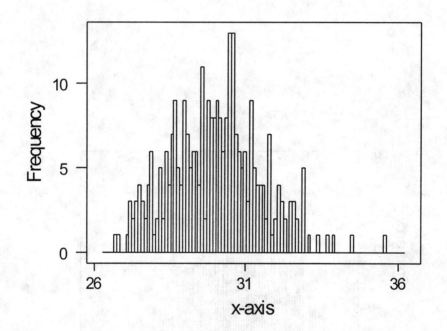

Sample Size = 30, 300 Samples

```
MTB > random 300 c1-c30;
SUBC> binomial n=100 p=.3.
MTB > rmean c1-c30 c35
MTB > Histogram c35;
SUBC>   MidPoint;
SUBC>   NInterval 100;
SUBC>   Bar;
SUBC>   Axis 1;
SUBC>     Label "x-axis";
SUBC>   Axis 2.
```

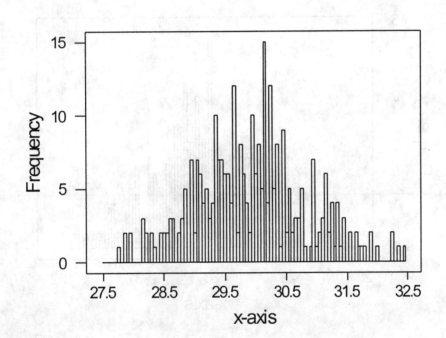

124   CHAPTER 8

Sample Size = 50, 300 Samples

```
MTB > random 300 c1-c50;
SUBC> binomial n=100 p=.3.
MTB > rmean c1-c50 c55
MTB > Histogram c55;
SUBC>   MidPoint;
SUBC>   NInterval 100;
SUBC>   Bar;
SUBC>   Axis 1;
SUBC>     Label "x-axis";
SUBC>   Axis 2.
```

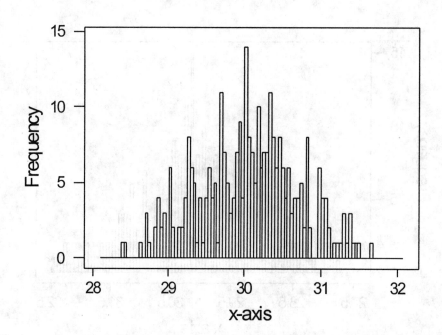

8.6 POISSON DISTRIBUTION

The *poisson* subcommand tells the *random* command to randomly generate numbers from the Poisson distribution. The argument of the *poisson* subcommand is the mean.

Sample Size = 5, 300 Samples

```
MTB > random 300 c1-c5;
SUBC> poisson 5.
MTB > rmean c1-c5 c10
MTB > Histogram c10;
SUBC>   MidPoint;
SUBC>   NInterval 100;
SUBC>   Bar;
SUBC>   Axis 1;
SUBC>     Label "x-axis";
SUBC>   Axis 2.
```

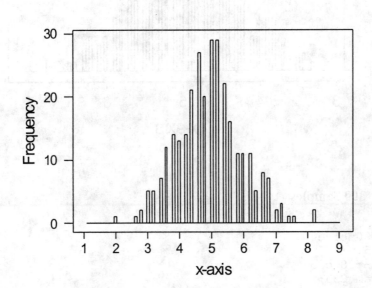

Sample Size = 10, 300 Samples

```
MTB > random 300 c1-c10;
SUBC> poisson 5.
MTB > rmean c1-c10 c15
MTB > Histogram C15;
SUBC>   MidPoint;
SUBC>   NInterval 100;
SUBC>   Bar;
SUBC>   Axis 1;
SUBC>     Label "x-axis";
SUBC>   Axis 2.
```

126   CHAPTER 8

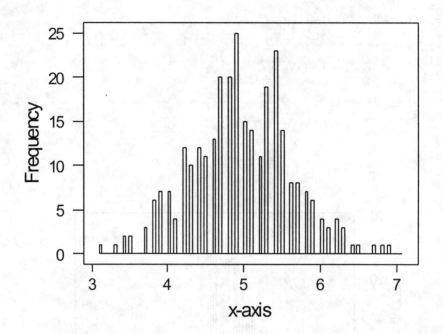

Sample Size = 30, 300 Samples

```
MTB > random 300 c1-c30;
SUBC> poisson 5.
MTB > rmean c1-c30 c35
MTB >  Histogram C35;
SUBC>    MidPoint;
SUBC>    NInterval 100;
SUBC>    Bar;
SUBC>    Axis 1;
SUBC>      Label "x-axis";
SUBC>    Axis 2.
```

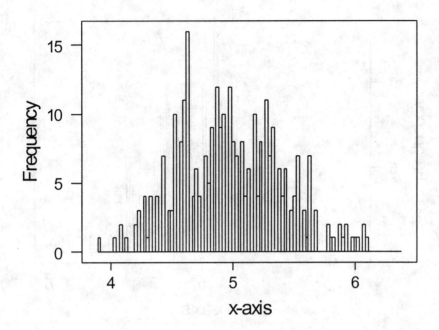

Sample Size = 50, 300 Samples

```
MTB > random 300 c1-c50;
SUBC> poisson 5.
MTB > rmean c1-c50 c55
MTB > Histogram c55;
SUBC>    MidPoint;
SUBC>    NInterval 100;
SUBC>    Bar;
SUBC>    Axis 1;
SUBC>       Label "x-axis";
SUBC>    Axis 2.
```

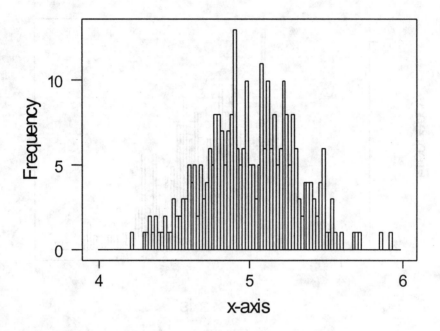

8.7 Statistical Summary

In this chapter we examined the important central limit theorem. This theorem states that the distribution of a mean of a sufficiently large sample can be approximated by a normal distribution no matter what kind of population distribution the sample came from. We demonstrated this theorem by randomly sampling from the distributions that we looked at in Chapters 6 and 7 and saw that each of these distributions can be approximated by a normal distribution when the sample size is sufficiently large.

CHAPTER 9
OTHER CONTINUOUS DISTRIBUTIONS
AND MOMENTS FOR DISTRIBUTIONS

9.1 Introduction

We will now look at other continuous distributions which are commonly used in statistics. It is important to study all these distributions because later we will make statistical inferences based on the particular distribution we are working with. We should remember that the area under the density curve of each of these distributions is equal to 1.

9.2 t Distribution

The t distribution is a symmetric distribution with mean 0.

The mathematical formula for the t distribution is

$$t = \frac{\overline{x} - \mu}{\frac{s_x}{\sqrt{n}}}$$

The t distribution is influenced by the degrees of freedom. As the degrees of freedom increase, the t distribution more closely resembles a standard normal distribution. In fact, the standard normal distribution is a t distribution with an infinite number of degrees of freedom. The characteristics of the t distribution are illustrated below.

```
MTB > Set c1
DATA>   1( -5 : 5 / .01 )1
DATA>   End.
MTB >  pdf c1 c2;
SUBC> t 1.
MTB > pdf c1 c3;
SUBC> t 4.
MTB > pdf c1 c4;
SUBC> normal 0 1.
MTB > name c1 'x-axis'
MTB > name c2 '1 deg'
MTB > name c3 '4 deg'
MTB > name c4 'std normal'
MTB > Plot c2*c1 c3*c1 c4*c1;
SUBC>    Connect;
SUBC>    Title "t Distribution and";
SUBC>    Title "Standard Normal Distribution";
SUBC>    Legend;
SUBC>      Type 1;
SUBC>      Color 0;
SUBC>    Overlay;
SUBC>    Axis 1;
SUBC>      Label "x-axis";
SUBC>    Axis 2;
SUBC>      Label "f(x)".
```

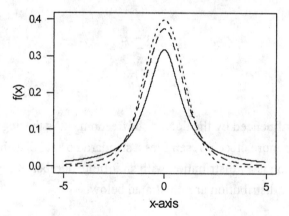

In the above graph the bottom distribution is a t distribution with one degree of freedom. The middle distribution is a t distribution with four degrees of freedom. The top distribution is a standard normal distribution, which has a mean of zero and a variance of one.

We will now calculate and plot the t cumulative density function to illustrate that the total probability of a t distribution equals 1.

```
MTB > Set c1
DATA>    1( -4 : 4 / .01 )1
DATA>    End.
MTB > cdf c1 c2;
SUBC> t 20.
MTB > Plot C2*C1;
SUBC>    Symbol;
SUBC>    Axis 1;
SUBC>       Label "x-axis";
SUBC>    Axis 2;
SUBC>       Label "f(x)".
MTB > Plot C2*C1;
SUBC>    Connect;
SUBC>    Axis 1;
SUBC>       Label "x-axis";
SUBC>    Axis 2;
SUBC>       Label "f(x)".
```

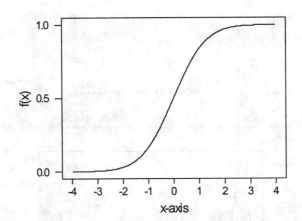

9.3 Chi-Square (χ^2) Distribution

Another distribution often used in statistics is the chi-square (χ^2) distribution. The mathematical formula for the chi-square is

$$\chi^2 = \sum_{i=1}^{n} (\frac{X_i - \mu}{\sigma_X})^2$$

This probability density function is greatly influenced by the degrees of freedom.

132   CHAPTER 9

The MINITAB session commands below will show that as the degree of freedom increases, the distribution shifts to the right. We will create three chi-square distributions with 5, 10, and 30 degrees of freedom to show that the chi-square distribution does in fact shift to the right as the degree of freedom increases.

```
MTB > Set c1
DATA>   1( 0 : 60 / .01 )1
DATA>   End.
MTB > pdf c1 c2;
SUBC> chisquare 5.
MTB > pdf c1 c3;
SUBC> chisquare 10.
MTB > pdf c1 c4;
SUBC> chisquare 30.
MTB > name c2 'df = 5'
MTB > name c3 'df = 10'
MTB > name c4 'df = 30'
MTB > Plot C2*C1 c3*c1 c4*c1;
SUBC>    Connect;
SUBC>       Type 1;
SUBC>       Color 1;
SUBC>       Size 1;
SUBC>    Overlay;
SUBC>    Axis 1;
SUBC>       Label "x-axis";
SUBC>    Axis 2;
SUBC>       Label "f(x)".
```

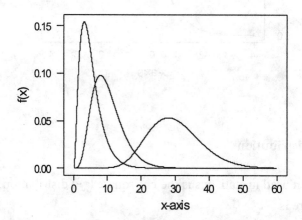

In the graph above, the left most distribution is a chi-square distribution with 5 degrees of freedom. The middle distribution is a chi-square distribution with 10 degrees of freedom. The right most distribution is a chi-square distribution with 30 degrees of freedom.

9.4 F Distribution

Another distribution often used in statistics is the F distribution. Its mathematical formula is

$$F = \frac{s_X^2 / \sigma_X^2}{s_Y^2 / \sigma_Y^2}$$

Each F distribution is influenced by two different degrees of freedom. There is a degree of freedom for the numerator and a degree of freedom for the denominator of the formula.

The session commands below will show that as the two different degrees of freedom increase, the more the F distribution shifts to the right and the more peaked the F distribution becomes. We will generate and plot three F distributions to illustrate this characteristic of the F distribution.

```
MTB > Set c1
DATA>    1( 0 : 5 / .01 )1
DATA>    End.
MTB > PDF c1 c2;
SUBC>    F 5 5.
MTB > pdf c1 c3;
SUBC> f 10 10.
MTB > pdf c1 c4;
SUBC> f 30 30.
MTB > Plot C2*C1 c3*c1 c4*c1;
SUBC>    Connect;
SUBC>       Type 1;
SUBC>       Color 1;
SUBC>       Size 1;
SUBC>    Overlay;
SUBC>    Axis 1;
SUBC>       Label "x-axis";
SUBC>    Axis 2;
SUBC>       Label "f(x)".
```

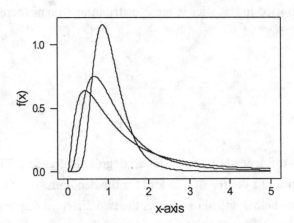

In the graph above, the leftmost distribution is an F distribution with 5,5 degrees of freedom. The middle distribution is an F distribution with 10,10 degrees of freedom. The rightmost distribution is an F distribution with 30,30 degrees of freedom.

9.5 Exponential Distribution

The last continuous distribution that we will look at is the exponential distribution. The formula for it in MINITAB is

$$f(t) = \frac{1}{\lambda} e^{-t/\lambda}, \quad t > 0$$

Some books use an alternative formula :

$$f(t) = \lambda e^{-\lambda t}, \quad t \geq 0$$

This second way of expressing the exponential formula has the interesting property that when $t = 0$ the exponential function will be equal to λ. The following session commands create exponential distributions with lambda equal to 1, 2, 3.

```
MTB > Set c1
DATA>   1( 0 : 4 / .01 )1
DATA>    End.
MTB > name c2 'lamda = 1'
MTB > name c3 'lamda = 2'
MTB > name c4 'lamda = 3'
MTB > pdf c1 c2;
SUBC> exponential 1.
MTB > pdf c1 c3;
SUBC> exponential 2.
MTB > pdf c1 c4;
SUBC> exponential 3.
MTB > Plot C2*C1 C3*C1 C4*C1;
SUBC>    Connect;
SUBC>    Overlay;
SUBC>    Axis 1;
SUBC>       Label "x-axis";
SUBC>    Axis 2;
SUBC>       Label "f(x)".
```

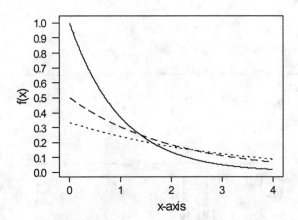

In the graph above the top distribution is where lambda equals to 1. The middle distribution is where lambda equals to 2. The bottom distribution is where lambda equals to 3.

136 CHAPTER 9

9.6 Central Limit Theorem

We will now illustrate the central limit theorem with the distributions we have examined in this chapter. As in Chapter 8, the distribution of the sample mean generated from the 4 distributions, which have been discussed in this chapter, will approach a normal distribution if the sample size is large enough.

Exponential Distribution

The *exponential* subcommand tells the *random* command to generate numbers at random from the exponential distribution. The argument of the exponential distribution indicates the λ of the exponential formula.

Sample Size = 5, 300 Samples

```
MTB > random 300 c1-c5;
SUBC> exponential 1.
MTB > rmean c1-c5 c10
MTB > Histogram C10;
SUBC>    MidPoint;
SUBC>    NInterval 100;
SUBC>    Bar;
SUBC>    Axis 1;
SUBC>      Label "x-axis";
SUBC>    Axis 2;
SUBC>      Label "Frequency".
```

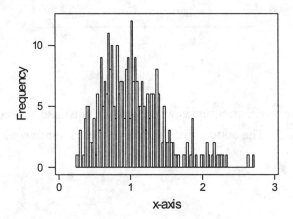

Sample Size = 10, 300 Samples

```
MTB >  random 300 c1-c10;
SUBC> exponential 1.
MTB > rmean c1-c10 c15
MTB > Histogram C15;
SUBC>   MidPoint;
SUBC>   NInterval 100;
SUBC>   Bar;
SUBC>   Axis 1;
SUBC>     Label "x-axis";
SUBC>   Axis 2;
SUBC>     Label "Frequency".
```

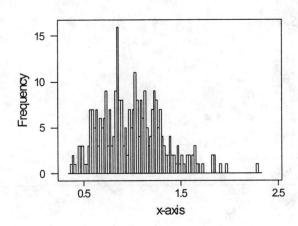

Sample Size = 30, 300 Samples

```
MTB > random 300 c1-c30;
SUBC> exponential 1.
MTB > rmean c1-c30 c35
MTB > Histogram C35;
SUBC>   MidPoint;
SUBC>   NInterval 100;
SUBC>   Bar;
SUBC>   Axis 1;
SUBC>     Label "x-axis";
SUBC>   Axis 2;
SUBC>     Label "Frequency".
```

138 CHAPTER 9

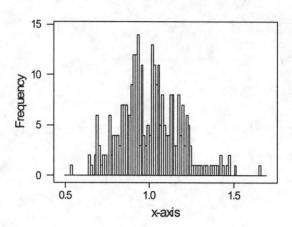

Sample Size = 50, 300 Samples

```
MTB > random 300 c1-c50;
SUBC> exponential 1.
MTB > rmean c1-c50 c55
MTB > Histogram C55;
SUBC>    MidPoint;
SUBC>    NInterval 100;
SUBC>    Bar;
SUBC>    Axis 1;
SUBC>       Label "x-axis";
SUBC>    Axis 2;
SUBC>       Label "Frequency".
```

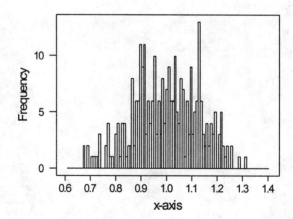

F Distribution

The *f* subcommand in MINITAB tells the *random* subcommand to generate numbers at random from the *F* distribution. The first argument of the *f* subcommand is the degree of freedom of the numerator. The second argument is the degree of freedom of the denominator.

Sample Size = 5, 300 Samples

```
MTB > random 300 c1-c5;
SUBC> f 5 5.
MTB > rmean c1-c5 c10
MTB > Histogram C10;
SUBC>    MidPoint;
SUBC>    NInterval 100;
SUBC>    Bar;
SUBC>    Axis 1;
SUBC>      Label "x-axis";
SUBC>    Axis 2;
SUBC>      Label "Frequency".
```

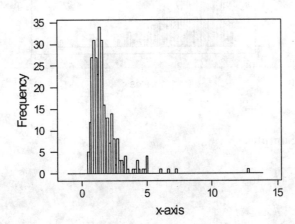

Sample Size = 10, 300 Samples

```
MTB > random 300 c1-c10;
SUBC> f 5 5.
MTB > rmean c1-c10 c15
MTB > Histogram C15;
SUBC>   MidPoint;
SUBC>   NInterval 100;
SUBC>   Bar;
SUBC>   Axis 1;
SUBC>     Label "x-axis";
SUBC>   Axis 2;
SUBC>     Label "Frequency".
```

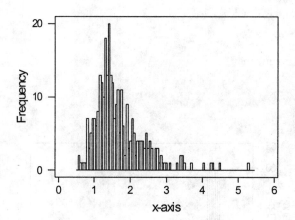

Sample Size = 30, 300 Samples

```
MTB >   random 300 c1-c30;
SUBC> f 5 5.
MTB > rmean c1-c30 c35
MTB > Histogram c35;
SUBC>   MidPoint;
SUBC>   NInterval 100;
SUBC>   Bar;
SUBC>   Axis 1;
SUBC>     Label "x-axis";
SUBC>   Axis 2;
SUBC>     Label "Frequency".
```

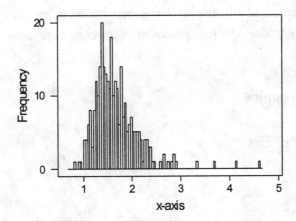

Sample Size = 50, 300 Samples

```
MTB > random 300 c1-c50;
SUBC> f 5 5.
MTB > rmean c1-c50 c55
MTB > Histogram C55;
SUBC>    MidPoint;
SUBC>    NInterval 100;
SUBC>    Bar;
SUBC>    Axis 1;
SUBC>      Label "x-axis";
SUBC>    Axis 2;
SUBC>      Label "Frequency".
```

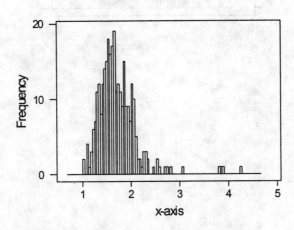

Chi-Square Distribution

The *chisquare* subcommand tells the *random* subcommand to generate numbers at random from the chi-square distribution. The first argument of the *chisquare* subcommand is degrees of freedom.

Sample Size = 5, 300 Samples

```
MTB >   random 300 c1-c5;
SUBC>   chisquare 5.
MTB >   rmean c1-c5 c10
MTB >   Histogram C10;
SUBC>     MidPoint;
SUBC>     NInterval 100;
SUBC>     Bar;
SUBC>     Axis 1;
SUBC>       Label "x-axis";
SUBC>     Axis 2;
SUBC>       Label "Frequency".
```

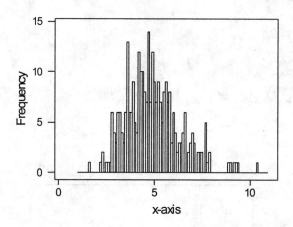

Sample Size = 10, 300 Samples

```
MTB > random 300 c1-c10;
SUBC> chisquare 5.
MTB > rmean c1-c10 c15
MTB > Histogram C15;
SUBC>   MidPoint;
SUBC>   NInterval 100;
SUBC>   Bar;
SUBC>   Axis 1;
SUBC>     Label "x-axis";
SUBC>   Axis 2;
SUBC>     Label "Frequency".
```

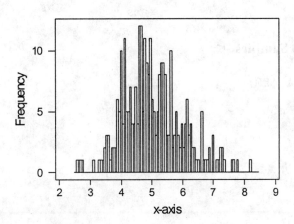

Sample Size = 30, 300 Samples

```
MTB > random 300 c1-c30;
SUBC> chisquare 5.
MTB > rmean c1-c30 c35
MTB > Histogram C35;
SUBC>   MidPoint;
SUBC>   NInterval 100;
SUBC>   Bar;
SUBC>   Axis 1;
SUBC>     Label "x-axis";
SUBC>   Axis 2;
SUBC>     Label "Frequency".
```

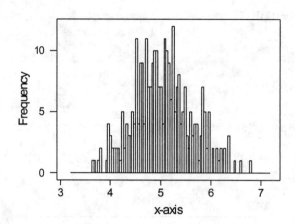

Sample Size = 50, 300 Samples

```
MTB > random 300 c1-c50;
SUBC> chisquare 5.
MTB > rmean c1-c50 c55
MTB > dotplot c55
```

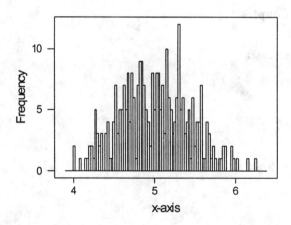

t Distribution

The *t* subcommand tells the *random* subcommand to generate numbers at random from the *t* distribution. The first argument of the *t* subcommand is the degree of freedom.

Sample Size = 5, 300 Samples

```
MTB >  random 300 c1-c5;
SUBC> t 10.
MTB > rmean c1-c5 c10
MTB > Histogram c10;
SUBC>    MidPoint;
SUBC>    NInterval 100;
SUBC>    Bar;
SUBC>    Axis 1;
SUBC>       Label "x-axis";
SUBC>    Axis 2;
SUBC>       Label "Frequency".
```

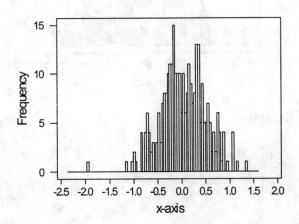

146   CHAPTER 9

Sample Size = 10, 300 Samples

```
MTB > random 300 c1-c10;
SUBC> t 10.
MTB > rmean c1-c10 c15
MTB > Histogram C15;
SUBC>   MidPoint;
SUBC>   NInterval 100;
SUBC>   Bar;
SUBC>   Axis 1;
SUBC>     Label "x-axis";
SUBC>   Axis 2;
SUBC>     Label "Frequency".
```

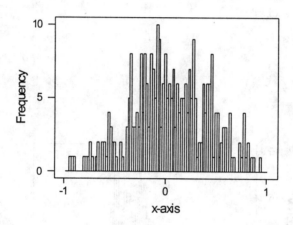

Sample Size = 30, 300 Samples

```
MTB > random 300 c1-c30;
SUBC> t 10.
MTB > rmean c1-c30 c35
MTB > Histogram C35;
SUBC>   MidPoint;
SUBC>   NInterval 100;
SUBC>   Bar;
SUBC>   Axis 1;
SUBC>     Label "x-axis";
SUBC>   Axis 2;
SUBC>     Label "Frequency".
```

OTHER CONTINUOUS DISTRIBUTIONS AND MOMENTS FOR DISTRIBUTIONS

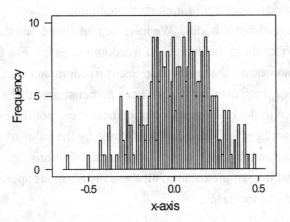

Sample Size = 50, 300 Samples

```
MTB > random 300 c1-c50;
SUBC> t 10.
MTB > rmean c1-c50 c55
MTB > Histogram C55;
SUBC>    MidPoint;
SUBC>    NInterval 100;
SUBC>    Bar;
SUBC>    Axis 1;
SUBC>       Label "x-axis";
SUBC>    Axis 2;
SUBC>       Label "Frequency".
```

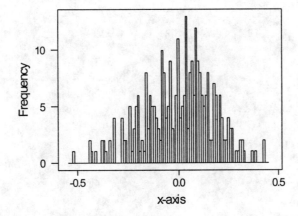

9.7 Statistical Summary

In this chapter we completed our survey of important continuous variables. We looked at the t, chi-square, F, and exponential distributions. We have demonstrated that the t distribution approaches a normal distribution as the degrees of freedom increase. For the chi-square distribution, we have demonstrated that as the degrees of freedom increase, the farther right the density function shifts. For the F distribution, we have demonstrated that as the degrees of freedom increase, the more peaked its density function becomes. For the exponential distribution, we have shown how the shape can be affected by the value of the parameter λ.

We also illustrated the central limit theorem with the distributions discussed in this chapter. We have demonstrated that the sample mean of all these distributions approaches a normal distribution as the sample size increases.

CHAPTER 10
ESTIMATION AND
STATISTICAL QUALITY CONTROL

10.1 Introduction

In this chapter we use the continuous variables which we studied in Chapters 7 and 9 to estimate the mean, variance, and proportion of populations.

Often we are interested in making inferences about a population parameter based on a sample parameter. One method for doing this is to make a confidence interval that might contain the population parameter. We are greatly aided in making this interval if we know the distribution of the sample.

Confidence intervals are intervals that have a chance of containing the population parameter. For example, a 90% confidence interval means that 90% of the confidence intervals that we create will contain the true population parameter. It is incorrect to say that a 90% confidence interval is an interval that has a 90% chance of containing the population parameter. The concept of confidence intervals is illustrated below.

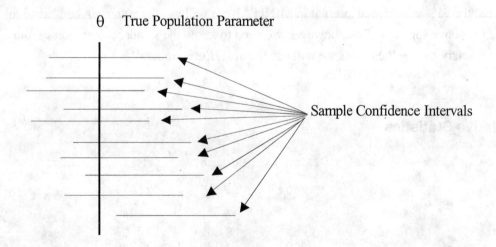

From this diagram of ten 90% confidence intervals, we can see that nine of the intervals contain the true population parameter while one of them does not contain the true mean.

10.2 Interval Estimates for μ when σ^2 is Known

In general, if the sample size is greater than 30, we know that the sample distribution will approximate a normal distribution. We can use this knowledge and the following formula to create the confidence interval for the population mean.

$$\bar{x} - z_{\alpha/2}(\sigma/\sqrt{n}) < \mu < \bar{x} + z_{\alpha/2}(\sigma/\sqrt{n})$$

Example 10.2A

Suppose we are interested in creating a 95% confidence interval for the following starting salaries for accountants.

```
$24,000  $20,000  $18,000  $26,000  $23,750  $32,000  $36,789  $25,300
$27,500  $35,000  $31,000  $25,500  $25,000  $50,000  $23,750  $30,500
$28,000  $29,000  $27,000  $27,750  $25,000  $41,000  $28,900  $32,000
$22,000  $26,000  $34,000  $26,550  $33,000  $27,900  $23,750  $31,000
$32,000  $21,430  $42,014  $36,450  $24,750  $34,000  $35,000  $27,000
```

We first need to put the data in column c1 of the worksheet.

To find the 95% confidence interval in MINITAB we will use the *zinterval* command to calculate the above formula. First, however, we need to know the standard deviation of our accounting salary data set. For this we will use the *describe* command.

```
MTB > describe c1
```

Descriptive Statistics

Variable	N	Mean	Median	TrMean	StDev	SE Mean
C1	40	29240	27825	28877	6352	1004

Variable	Minimum	Maximum	Q1	Q3
C1	18000	50000	25000	32750

```
MTB > zinterval 95 6352 c1
```

Z Confidence Intervals

```
The assumed sigma = 6352

Variable      N      Mean    StDev   SE Mean      95.0 % CI
C1           40     29240     6352      1004  (  27271,    31208)
```

The *describe* command describes the data in column c1 using several descriptive statistical measures. We need to use the *describe* command because we need to know the standard deviation of the data set for the *zinterval* command. Here the standard deviation is shown as 6352.

The *zinterval* command is used to calculate a confidence interval when σ is known. The first argument of this command, '95', indicates that we are interested in a 95 % confidence interval. The second argument, '6352', is the standard deviation of the data set. We got this information by issuing the *describe* command. The last argument, 'c1', is the location of the data set.

MINITAB calculated that the 95% confidence interval for the accounting salaries is between $27,271 and $31,208 and that the mean is $29,240.

Often we are given the sample size, mean, and standard deviation but not the data set. Without the data set we cannot use the *zinterval* command to create the confidence interval. To solve this problem, we will create a macro to calculate the confidence interval. The macro will be called *zint*.

Zint Macro

```
gmacro
noecho
#
#  This macro is used  chapter 10
note calculates a mean confidence interval using a normal distribution
note What is the mean?
set 'terminal' c50;
nobs=1.
End

note What is the standard deviation?
set 'terminal' c51;
nobs=1.
End

note What is the sample size of the data set?
set 'terminal' c52;
nobs=1.
End
```

```
note What is the desired percentage confidence interval?
set 'terminal' c53;
nobs=1.
End

let k50=c50
#k50 has the mean number
let k51=c51
#k51 has the standard deviation
let k52=c52
#k52 has the sample size of the data set
let k53=c53
#k53 has the confidence interval percentage
let k55=(1+k53)/2
invcdf k55 k56;
normal 0 1.

#calculating the upper and lower end of interval
let c61=k50+k56*(k51/sqrt(k52))
let c60=k50-k56*(k51/sqrt(k52))

name c60 'lowint' c61 'highint'
note The confidence interval is:
print c60 c61

erase c60-c70
endmacro
```

Example 10.2B

Suppose a real estate agent in Connecticut is interested in the mean home price in the state. A random sample of 50 homes shows a mean price of $175,622, assuming a population sample standard deviation of $37,221. What is the 95 % confidence interval for the mean home price for Connecticut?

```
MTB > %zint
Executing from file: zint.MAC
calculates a mean confidence interval using a normal distribution
What is the mean?
DATA> 175622
What is the standard deviation?
DATA> 37221
What is the sample size of the data set?
DATA> 50
What is the desired percentage confidence interval?
DATA> .95
The confidence interval is:
```

Data Display

```
Row    lowint    highint
  1    165305    185939
```

Our output tells us that the 95% confidence interval for the mean home price in Connecticut is from $165,305 to $185,939.

This is graphically illustrated below.

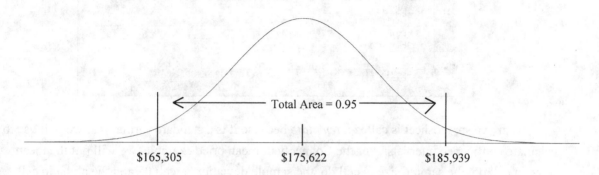

In MINITAB there are two ways to put comments in a macro. The first way is to use the *note* command. The second way is to use the # command. The *noecho* command will display the comments of the *note* command when the macro is executed. The *noecho* command will not display the comments of the # command when the macro is executed.

In EXCEL we will use the *confidence* function to calculate the confidence interval for the mean home price. The *confidence* function has three parameters. They are,

Alpha: The significance level used to compute the confidence level.
Standard_dev: The standard deviation for the data range.
Size : The sample size.

How to setup an Excel spreadsheet to calculate the confidence interval is shown below.

154   CHAPTER 10

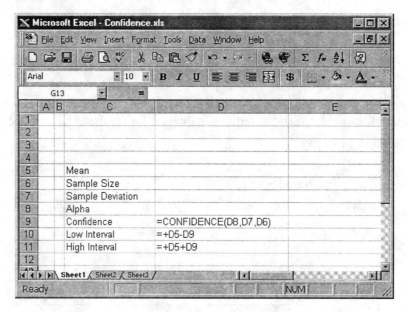

The above spreadsheet is called a *template* because it is a standard format that we will use to calculate confidence intervals. For the Connecticut mean price example, we will put the mean price in cell d5, the sample size in cell d6, the sample deviation in cell d7 and the alpha in cell d8. This is shown below.

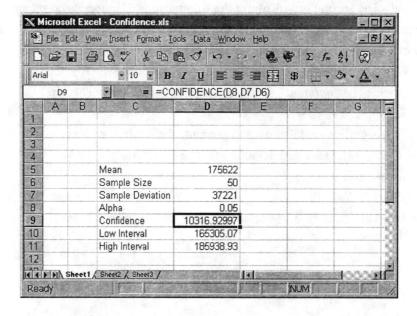

Example 10.2C

Suppose a machine dispenses sand into bags. The population standard deviation is 9.0 pounds, and the weights are normally distributed. A random sample of 100 bags is taken, and the sample mean is 105 pounds. Calculate a 95% confidence interval for the mean weight of the bags using the *zint* macro.

```
MTB > %zint
Executing from file: zint.MAC
calculates a mean confidence interval using a normal distribution
What is the mean?
DATA> 105
What is the standard deviation?
DATA> 9
What is the sample size of the data set?
DATA> 100
What is the desired percentage confidence interval?
DATA> .95
The confidence interval is:
```

Data Display

```
Row    lowint    highint

 1     103.236   106.764
```

From the output we can see that the 95% confidence interval for the mean weight of the bags is 103.236 pounds to 106.764 pounds.

This is graphically illustrated below.

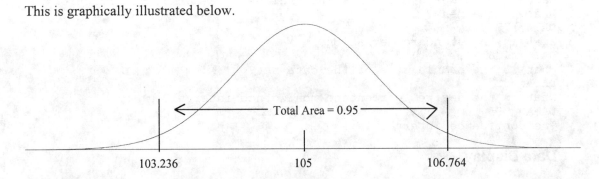

The calculation in EXCEL is shown below.

Example 10.2D

To study the effect of internal audit departments on external audit fees, W.A. Wallace conducted a survey of the audit department of 32 diverse companies (*Harvard Business Review*, March-April 1984). She found that the mean annual external audit paid by the 32 companies was $779,030 and the standard deviation was $1,083,162. Use MINITAB to find the 95% confidence interval for the mean external audit fees.

```
MTB > %zint
Executing from file: zint.MAC
calculates a mean confidence interval using a normal distribution
What is the mean?
DATA> 779030
What is the standard deviation?
DATA> 1083162
What is the sample size of the data set?
DATA> 32
What is the desired percentage confidence interval?
DATA> .95
The confidence interval is:
```

Data Display

```
Row    lowint     highint

 1     403740    1154320
```

From the *zint* macro output we can see that the 95% confidence interval for the mean external audit fees is from $403,740 to $1,154,320.

This is graphically illustrated below.

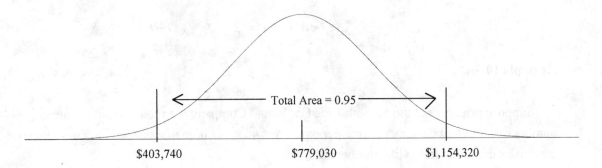

The calculation in EXCEL is shown below.

Mean	779030	
Sample Size	32	
Sample Deviation	1083162	
Alpha	0.05	
Confidence	375289.0338	
Low Interval	403740.9662	
High Interval	1154319.034	

Formula: =CONFIDENCE(D8,D7,D6)

158   CHAPTER 10

10.3 Confidence Intervals for μ When σ^2 is Unknown

When the sample size is small and the data have a normal distribution, we will have to use the *t* test. The *t* test uses a *t* distribution. The formula to create the confidence interval would be:

$$\bar{x} - t_{n-1,\alpha/2}(s/\sqrt{n}) < \mu < \bar{x} + t_{n-1,\alpha/2}(s/\sqrt{n})$$

Example 10.3A

Suppose managers at the Smooth Ride Car Rental Company are interested in the mean number of miles that customers drive per day. A random sample of 6 car rentals shows that the customers drove the following numbers of miles:

152, 222, 300, 84, 90, 122

Construct a 99 % confidence interval for the mean number of miles driven.

```
MTB > tinterval 99 c1
```

T Confidence Intervals

```
Variable    N      Mean    StDev   SE Mean       99.0 % CI
C1          6     161.7     84.4      34.5   (   22.7,   300.6)
```

We can see from the MINITAB output that the 99% confidence interval is between 22.7 miles and 300.6 miles.

We will now calculate the interval in EXCEL. Instead of using the *confidence* function we will use the *tinterval* function.

Example 10.3B

Suppose that in the above example the managers did not give us the data set and instead gave us the mean and standard deviation. If this was the case then we could not use the *tinterval* command. Instead, we would have to create our own macro to calculate the confidence interval. We will call this macro *tint*. It turns out that we can use most of the *zint* macro commands. The *tint* macro is shown below.

Tint **Macro**

```
Gmacro
noecho
#  This macro is used in chapter 10
note calculates a mean confidence interval using a t distribution
note What is the mean?
set 'terminal' c50;
nobs=1.
End

note What is the standard deviation?
set 'terminal' c51;
nobs=1.
End

note What is the sample size of the data set?
set 'terminal' c52;
nobs=1.
End

note What is the desired percentage confidence interval?
set 'terminal' c53;
nobs=1.
End

let k50=c50
#k50 has the mean number
let k51=c51
#k51 has the standard deviation
let k52=c52
#k52 has the sample size of the data set
let k53=c53
#k53 has the confidence interval percentage

let k55=(1+k53)/2
let k57=k52-1
# k57 has the degree of freedom for t
invcdf k55 k56;
t k57.

# t distribution with k57 degree of freedom
let c61=k50+k56*(k51/sqrt(k52))
let c60=k50-k56*(k51/sqrt(k52))
name c60 'lowint' c61 'highint'

note The confidence interval is:
print c60 c61
erase c60-c70
endmacro
```

Example 10.3C

Suppose after randomly selecting 8 football players, we found that the mean weight was 212.75 and the sample standard deviation was 23.34. Using the *tint* macro, calculate the 95% confidence interval.

```
MTB > %tint
Executing from file: tint.MAC
calculates a mean confidence interval using a t distribution
What is the mean?
DATA> 212.75
What is the standard deviation?
DATA> 23.34
What is the sample size of the data set?
DATA> 8
What is the desired percentage confidence interval?
DATA> .95
The confidence interval is:
```

Data Display

```
Row    lowint    highint

 1    193.237    232.263
```

The output then indicates that the 95% confidence interval for the average weight of football players is from 193.237 pounds to 232.263 pounds.

This is illustrated below.

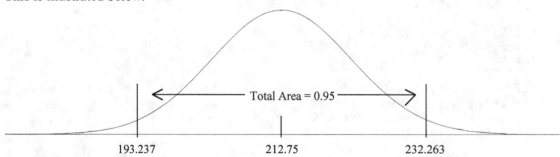

10.4 Confidence Intervals for the Population Proportion

Because of the central limit theorem, whenever we calculate the confidence interval for the population proportion, we will use the normal distribution if the sample is large enough. The formula to calculate the confidence interval for the population proportion is

$$\hat{P} - z_{\alpha/2}\sqrt{\frac{\hat{P}(1-\hat{P})}{n}} < P < \hat{P} + z_{\alpha/2}\sqrt{\frac{\hat{P}(1-\hat{P})}{n}}$$

Since MINITAB does not have a command for the confidence interval for the population proportion, we have to create our own. We will call our macro *pint*.

Example 10.4A

A marketing firm discovers that 65% of the 30 customers who participated in a blind taste test prefer brand A to brand B. Develop a 95% confidence interval for the number of people who prefer brand A using the *pint* macro.

Pint Macro

```
gmacro
noecho
#
#  This macro is used in chapter 10
note calculates a proportion confidence interval using a normal distribution
note What is the proportion?
set 'terminal' c50;
nobs=1.
End

note What is the sample size of the data set?
set 'terminal' c52;
nobs=1.
End

note What is the desired percentage confidence interval?
set 'terminal' c53;
nobs=1.
End

let k50=c50
#k50 has the mean number
let k52=c52
#k52 has the sample size of the data set
```

```
let k53=c53
#k53 has the confidence interval percentage
let k55=(1+k53)/2
invcdf k55 k56;
normal 0 1.
let c61=k50+k56*(sqrt((k50*(1-k50))/k52))
let c60=k50-k56*(sqrt((k50*(1-k50))/k52))
name c60 'lowint' c61 'highint'
note The confidence interval is:
print c60 c61
erase c60-c70
endmacro



MTB > %pint
Executing from file: pint.MAC
calculates a proportion confidence interval using a normal distribution
What is the proportion?
DATA> .65
What is the sample size of the data set?
DATA> 30
What is the desired percentage confidence interval?
DATA> .95
The confidence interval is:
```

Data Display

Row	lowint	highint
1	0.479322	0.820678

From the *pint* macro the confidence interval for those who preferred brand A lies between 47.9322 % and 82.0678 %.

Graphically the confidence interval would look like the following.

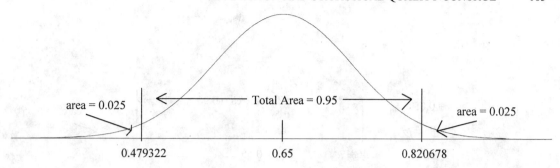

Example 10.4B

Suppose that a random sample of 100 voters is taken, and 55% of the sample supports the incumbent candidate. Construct a 90% confidence interval for the proportion that supports the incumbent candidate.

```
MTB > %pint
Executing from file: pint.MAC
calculates a proportion confidence interval using a normal distribution
What is the proportion?
DATA> .55
What is the sample size of the data set?
DATA> 100
What is the desired percentage confidence interval?
DATA> .90
The confidence interval is:
```

Data Display

```
Row     lowint      highint

 1     0.468170    0.631830
```

From the *pint* macro we learn that the confidence interval for those who support the incumbent candidate is from 46.817 % to 63.183 %.

Graphically the confidence interval looks like the following.

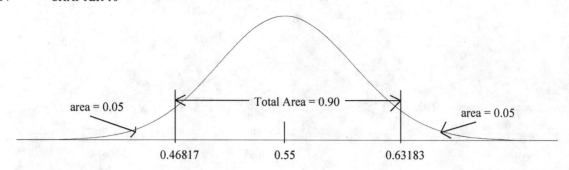

Example 10.4C

A study in the *Journal of Advertising Research* (April/May 1984) to find the proportion of working adults using computer equipment on the job employed the random sample approach to survey 616 working adults. The survey revealed that 184 of the adults, or 29.9%, regularly used computer equipment on the job. Use the *pint* macro to find the 95% confidence interval for working adults' computer usage.

```
MTB > %pint
Executing from file: pint.MAC
calculates a proportion confidence interval using a normal distribution
What is the proportion?
DATA> .299
What is the sample size of the data set?
DATA> 616
What is the desired percentage confidence interval?
DATA> .95
The confidence interval is:
```

Data Display

```
Row     lowint      highint

 1     0.262846    0.335154
```

From the *pint* macro we can see that the 95% confidence interval for working adults' computer usage was from 26.2846 % to 33.5154 %.

This is graphically shown below.

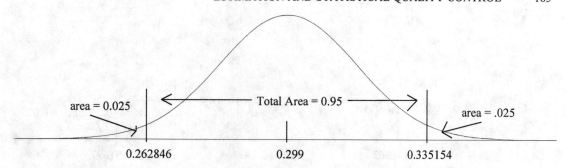

Example 10.4D

Guffey, Harris, and Laumer (1979) studied the attitudes of shoppers toward shoplifting and devices for its prevention. They sampled 403 shopping center patrons. Twenty-four percent of this sample expressed awareness of and discomfort with the use of TV cameras as a device to prevent shoplifting. Use the *pint* macro to find the 95% confidence interval of the population in order to find the proportion that dislikes the use of TV devices to prevent shoplifting.

```
MTB > %pint
Executing from file: pint.MAC
calculates a proportion confidence interval using a normal distribution
What is the proportion?
DATA> .24
What is the sample size of the data set?
DATA> 403
What is the desired percentage confidence interval?
DATA> .95
The confidence interval is:
```

Data Display

```
 Row    lowint    highint

   1   0.198303   0.281697
```

From the *pint* macro we can see that the 95% confidence interval proportion that dislikes using TV cameras to prevent shoplifting is from 19.8303 % to 28.1697 %.

CHAPTER 10

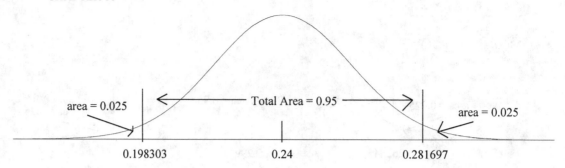

10.5 Confidence Intervals for the Variance

The confidence interval for variances uses a chi-square distribution. The formula for the confidence interval for a variance is:

$$\frac{(n-1)s_x^2}{\chi^2_{v,\alpha/2}} < \sigma^2 < \frac{(n-1)s_x^2}{\chi^2_{v,1-\alpha/2}}$$

Since MINITAB does not have a command for the confidence interval for the variance, we are going to have to create our own macro called *xint*.

Suppose a random sample of 30 bags of sand is taken and the sample variance of weights is 5.5. Find a 95% confidence interval for the population variance using the *xint* macro.

Xint Macro

```
gmacro
noecho
#  This macro is used in chapter 10
note calculates a variance confidence interval using chisquare
distribution
note What is the sample variance?
set 'terminal' c51;
nobs=1.
End

note What is the sample size of the data set?
set 'terminal' c52;
nobs=1.
End

note What is the desired percentage confidence interval?
set 'terminal' c53;
nobs=1.
End
```

```
let k51=c51
#k51 has the sample variance
let k52=c52
#k52 has the sample size of the data set
let k53=c53
#k53 has the confidence interval percentage

let k60=k52-1
let k55=(1+k53)/2
invcdf k55 k56;
chisquare k60.
let c60=((k60)*(k51))/k56
let k61=1-k55
invcdf  k61 k57;
chisquare k60.
let c61=((k60)*(k51))/k57
name c60 'lowint' c61 'highint'
note The confidence interval is:
print c60 c61
erase c60-c70
endmacro


MTB > %xint
Executing from file: xint.MAC
calculates a variance confidence interval using chisquare distribution
What is the sample variance?
DATA> 5.5
What is the sample size of the data set?
DATA> 30
What is the desired percentage confidence interval?
DATA> .95
The confidence interval is:
```

Data Display

Row	lowint	highint
1	3.48845	9.93951

From the *xint* macro we learn that the confidence interval for the sample variance of sandbag weight is between 3.4885 pounds and 9.93951 pounds.

10.6 Statistical Summary

In this chapter we created the confidence interval for the population mean when the population variance was known and unknown. We also found the confidence interval for the population proportion and the confidence interval for the population variance.

CHAPTER 11
HYPOTHESIS TESTING

11.1 Introduction

In the last chapter we made inferences about a population parameter by creating a confidence interval from a sample. We will now look at another method, called *hypothesis testing*, for making inferences about a population parameter. In hypothesis testing we infer that the stated null hypothesis (H_0) is true until there is convincing but *not perfect* evidence that the null hypothesis is not true. Our evidence is from the sample that we obtain. We conclude that there is convincing evidence when the *p*-value is less than the alpha value. The *p*-value will be discussed later in this chapter.

Because the sample evidence is not perfect, we can make four kinds of decisions.

1. We can make a correct decision by rejecting H_0 when in fact the true H_0 is false.
2. We can make a correct decision by failing to reject H_0 when in fact the true H_0 is true.
3. We can make an incorrect decision by rejecting H_0 when in fact the true H_0 is true.
4. We can make an incorrect decision by accepting H_0 when in fact the true H_0 is false.

In statistics we call situation 3 a Type I error.
We will make a Type I error alpha percent of the time.
In statistics we call situation 4 a Type II error.
Below is a diagram of the four possible resulting decisions of hypothesis testing.

	Actual H_0 True	Actual H_0 False
Fail to Reject H_0	Correct Decision	Type II Error
Reject H_0	Type I Error	Correct Decision

It is interesting to note that some statisticians feel that a jury in a court of law is a situation in which Type I and Type II errors might be made.

11.2 One-Tailed Tests of Mean for Large Samples

We begin by examining the case where only one sample is drawn and that sample is large. Using a large sample offers two important advantages. First, we can apply the central limit theorem. Second, the large sample enables us, through our choice of significance level, alpha, to reduce our chance of making a Type II error.

Example 11.2A

Suppose an accountant is interested in finding out if the average starting salary for accountants is greater than $25,000 or not. Test at the 5% alpha level. The accountant collected the following sample of 40 salaries.

$24,000 $20,000 $18,000 $26,000 $23,750 $32,000 $36,789 $25,300
$27,500 $35,000 $31,000 $25,500 $25,000 $50,000 $23,750 $30,500
$28,000 $29,000 $27,000 $27,750 $25,000 $41,000 $28,900 $32,000
$22,000 $26,000 $34,000 $26,550 $33,000 $27,900 $23,750 $31,000
$32,000 $21,430 $42,014 $36,450 $24,750 $34,000 $35,000 $27,000

The null and alternative hypotheses would then be

$$H_0 = \$25,000$$
$$H_1 > \$25,000$$

The test statistic for this would be

$$\frac{\bar{x} - \mu_0}{\sigma/\sqrt{n}} > Z_\alpha$$

To do this test we should first put the salary data in column c1 of the MINITAB worksheet and then we will use the *ztest* command to perform the calculation.

```
MTB > set c1
DATA> 24000 20000 18000 26000 23750 32000 36789 25300
DATA> 27500 35000 31000 25500 25000 50000 23750 30500
DATA> 28000 29000 27000 27750 25000 41000 28900 32000
DATA> 22000 26000 34000 26550 33000 27900 23750 31000
DATA> 32000 21430 42014 36450 24750 34000 35000 27000
DATA> end
MTB > describe c1
```

Descriptive Statistics

```
Variable              N        Mean      Median     TrMean      StDev     SE Mean
C1                   40       29240       27825      28877       6352        1004

Variable        Minimum     Maximum          Q1         Q3
C1                18000       50000       25000      32750
```

MTB > ztest 25000 6352 c1;
SUBC> alternate 1.

Z-Test

```
Test of mu = 25000 vs mu > 25000
The assumed sigma = 6352

Variable        N        Mean       StDev    SE Mean          Z          P
C1             40       29240        6352       1004       4.22     0.0000
```

MTB > invcdf .95;
SUBC> normal 0 1.

Inverse Cumulative Distribution Function

Normal with mean = 0 and standard deviation = 1.00000

```
P( X <= x)         x
   0.9500       1.6449
```

From the *ztest* command output the calculated sample Z score for the data set is 4.22. We are testing at a 5% alpha level. For a one-sided right-tailed test, the 5% alpha level begins at the value 1.6449. We will reject the null hypothesis when the calculated Z value is greater than 1.6449 or when the calculated Z value lies in the 5% alpha area. The test statistic which we calculated, 4.22, lies in the reject region.

Graphically we are seeing the following:

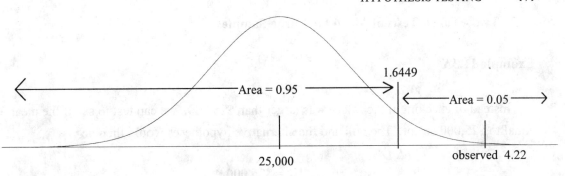

Hypothesis Testing and the *p*-value

Another way to test the null hypothesis is to use the *p*-value. A *p*-value is the probability of getting a value more extreme than the calculated statistic. If we see that the *p*-value is less than the alpha value, we know that there is significant evidence to reject the null hypothesis because the *p*-value would lie in the region of the alpha value as shown in the above graph. In our example, the *p*-value is the area to the right of the calculated value of 4.22. This area to the right of 4.22 is the probability of getting a value more extreme than the calculated statistic.

With the advent of computer statistical software, the use of the *p*-value method for accepting or rejecting the null hypothesis is easier than using the test statistic method. In the latter method we would also have to look up the value that begins the alpha value. In our example the value 1.6449 is the value where an alpha value of 5% begins. We then have to see if the calculated value is greater than the value of 1.6449. If it is, we conclude that there is enough evidence to reject the null hypothesis. If not, we conclude there is not enough evidence to reject the null hypothesis.

In the *p*-value method we only need to see if the *p*-value is greater or less than the alpha. If it is greater or equal to the alpha value, we accept the null hypothesis. If it is less than the alpha value, we reject the null hypothesis.

The calculated *p*-value in our example is 0.00. Since this *p*-value is less than the alpha value of 5%, there is enough evidence to reject the null hypothesis that the mean accounting salary is $25,000 and accept the alternative hypothesis that the mean accounting salary is greater than $25,000.

We should remember that we have strong but not perfect evidence to reject the null hypothesis. There is a 5% chance that we should not be rejecting the null hypothesis. As stated above, this kind of error in judgment is called a Type I error.

172   CHAPTER 11

11.3 Two - Tailed Tests of Mean for Large Samples

Example 11.3A

Instead of checking if the salary was larger than $25,000, we can test to see if the mean is equal to $25,000 or not. The null and the alternative hypotheses would then be

$$H_0 = \$25,000$$
$$H_1 \neq \$25,000$$

This would represent a two-tailed test. The mathematical formula to calculate this would be

$$\frac{\bar{x} - \mu_0}{\sigma/\sqrt{n}} > Z_{\alpha/2} \quad \text{or} \quad \frac{\bar{x} - \mu_0}{\sigma/\sqrt{n}} < -Z_{\alpha/2}$$

The following shows how MINITAB would test the hypothesis.

```
MTB > describe c1
```

Descriptive Statistics

```
Variable         N       Mean     Median     TrMean      StDev    SE Mean
C1              40      29240      27825      28877       6352       1004

Variable    Minimum    Maximum         Q1         Q3
C1            18000      50000      25000      32750

MTB > ztest 25000 6352 c1
```

Z-Test

```
Test of mu = 25000 vs mu not = 25000
The assumed sigma = 6352

Variable     N       Mean      StDev    SE Mean          Z          P
C1          40      29240       6352       1004       4.22     0.0000
```

```
MTB > invcdf .975;
SUBC> normal 0 1.
```

Inverse Cumulative Distribution Function

```
Normal with mean = 0 and standard deviation = 1.00000

P( X <= x)         x
   0.9750       1.9600
```

From the *ztest* command output the calculated Z score for the data set is 4.22. We will reject the null hypothesis if the sample Z score is greater than the standard Z score of 1.96. Since this is the case, we can reject the null hypothesis that the mean salary is $25,000 and conclude that the mean salary is different from $25,000.

Using the *p*-value method, we can see that the calculated *p*-value is 0.0. Since this is less than the alpha value of 0.05, indicating that the *p*-value is in the alpha area, we would reject the null hypothesis and accept the alternative hypothesis.

Graphically we are seeing the following:

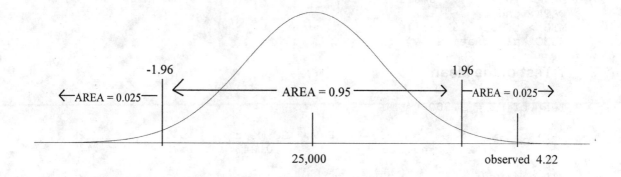

11.4 One-Tailed Tests of Mean for Small Samples

For one-tailed tests of mean for samples of less than 30, the *t* test should be used. The *t* test, which uses a *t* distribution, looks like the following.

$$\frac{\bar{x} - \mu_0}{s/\sqrt{n}} = t$$

We will use the *t* test in the following example.

CHAPTER 11

United Van Lines Company is considering purchasing a large, new moving van. The sales agency has agreed to lease the truck to United Van Lines for 4 weeks (24 working days) on a trial basis. The main concern of United Van Lines is the miles per gallon (mpg) of gasoline that the van obtains on a typical moving day. The mpg values for the 24 trial days are

$$8.5 \quad 9.5 \quad 8.7 \quad 8.9 \quad 9.1 \quad 10.1 \quad 12 \quad 11.5 \quad 10.5 \quad 9.6$$
$$8.7 \quad 11.6 \quad 10.9 \quad 9.8 \quad 8.8 \quad 8.6 \quad 9.4 \quad 10.8 \quad 12.3 \quad 11.1$$
$$10.2 \quad 9.7 \quad 9.8 \quad 8.1$$

They want to know if the average was greater than 9.5. Our hypothesis would then be

$$H_0 = 9.5$$
$$H_1 > 9.5$$

```
MTB > set c1
DATA> 8.5 9.5 8.7 8.9 9.1 10.1 12 11.5 10.5 9.6
DATA> 8.7 11.6 10.9 9.8 8.8 8.6 9.4 10.8 12.3 11.1
DATA> 10.2 9.7 9.8 8.1
DATA> end
MTB > ttest 9.5 c1;
SUBC> alternative 1.
```

T-Test of the Mean

```
Test of mu = 9.500 vs mu > 9.500

Variable      N      Mean    StDev   SE Mean       T        P
C1           24     9.925    1.189     0.243    1.75    0.047

MTB > invcdf .95;
SUBC> t 23.
```

Inverse Cumulative Distribution Function

```
Student's t distribution with 23 DF

 P( X <= x)           x
   0.9500         1.7139
```

The critical value for 95% was 1.7109. The calculated value for the sample was 1.75. Since the calculated value was larger than the critical value, we reject the null hypothesis and accept the alternative hypothesis.

Using the *p*-value method we can see that the calculated *p*-value of 0.047 is less than the alpha value of 0.05. Therefore we can conclude that there is enough evidence to reject the null hypothesis and accept the alternative hypothesis.

Graphically this situation would look like the following:

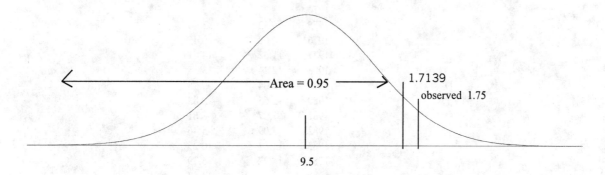

11.5 Difference of Two Means

Often we are interested in knowing if there is any difference between two population means. To illustrate this concept we will test to see if the average EPS for GM and Ford are equal or not.

Example 11.5A

The hypotheses to be tested are

$$H_0: \mu_{GM} - \mu_{FORD} = 0$$
$$H_1: \mu_{GM} - \mu_{FORD} \neq 0$$

The data for the EPS for GM and Ford are shown below.

EPS FOR GM AND FORD

YEAR	GM	FORD
1970	2.09	4.77
1971	6.72	6.18
1972	7.51	8.52
1973	8.34	9.13
1974	3.27	3.86
1975	4.32	3.46
1976	10.08	10.45
1977	11.62	14.61
1978	12.24	13.35
1979	10.04	9.75
1980	-2.65	-12.83
1981	1.07	-8.81
1982	3.09	-5.46
1983	11.84	10.29
1984	14.22	15.79
1985	12.28	13.63
1986	8.22	12.32
1987	10.06	18.10
1988	13.64	10.96
1989	6.33	8.22
1990	-4.09	1.86

To do this analysis, we will put the GM data in column c1 in the MINITAB worksheet and the Ford data in column c2. We will also name column c1 *'gm'* and column c2 *'ford'*.

When testing to see if the samples of two means are equal or not, it is helpful to graph the data. In this case we will dotplot the GM and Ford data, as shown below.

```
MTB > name c1 'GM'
MTB > name c2 'FORD'
MTB > set 'GM'
DATA> 2.09 6.72 7.51 8.34 3.27 4.32 10.08 11.62 12.24 10.04
DATA> -2.65 1.07 3.09 11.84 14.22 12.28 8.22 10.06 13.64 6.33 -4.09
DATA> end
MTB > set 'FORD'
DATA> 4.77 6.18 8.52 9.13 3.86 3.46 10.45 14.61 13.35 9.75
DATA> -12.83 -8.81 -5.46 10.29 15.79 13.63 12.32 18.10 10.96 8.22 1.86
DATA> END
MTB > dotplot c1 c2;
SUBC> same.
```

Dotplot

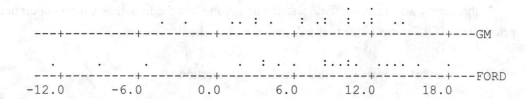

The *same* subcommand tells the *dotplot* command to use the same scale for both plots.

The resulting dotplots show that there is visually not much difference between the mean EPS of GM and that of Ford. Now let us use MINITAB to mathematically calculate whether there is a significant difference.

```
MTB > twosample .95 c1 c2;
SUBC> pooled.
```

Two Sample T-Test and Confidence Interval

```
Two sample T for GM vs FORD

         N      Mean    StDev   SE Mean
GM      21      7.15     5.19      1.1
FORD    21      7.05     7.99      1.7

95% CI for mu GM - mu FORD: ( -4.1,   4.3)
T-Test mu GM = mu FORD (vs not =): T = 0.05   P = 0.96   DF = 40
Both use Pooled StDev = 6.74
```

We use the *twosample* command to see if the two means are equal or not. The first argument, '.95', of the *twosample* command is the confidence level we want. The second argument, 'c1', is the location of the data in the first sample. The third argument, 'c2', is the location of the second data set. The *pooled* subcommand lets the *twosample* command assume that the two populations have equal variances.

The calculated *p*-value is 0.96. So at the 0.05 alpha level, we cannot reject the hypothesis that the average EPS for GM and Ford are equal.

11.6 Hypothesis Testing for A Population Proportion

If the sample size is large, the Z test can be used to test the hypothesis for a population proportion. If the sample size is small, then the t test should be used.

The test statistic for a two-tailed population proportion is

$$-z_{\alpha/2} > \frac{\hat{P} - P_0}{\sqrt{P_0(1-P_0)/n}} \text{ or } z_{\alpha/2} > \frac{\hat{P} - P_0}{\sqrt{P_0(1-P_0)/n}}$$

Example 11.6A

Francis Company is evaluating the promotability of its employees. That is, it is determining the proportion of employees whose ability, training, and supervisory experience qualify them for promotion to the next level of management. The human resources director of Francis Company tells the president that 80% of the employees in the company are 'promotable.' However, a special committee appointed by the president finds that only 75% of the 200 employees who have been interviewed are qualified for promotion. Use this information to do a two-tailed hypothesis test at $\alpha = 5\%$.

The null and alternative hypotheses would be

$$H_0: p = 0.80$$
$$H_1: p \neq 0.80$$

MINITAB does not have a command to do a two-tailed hypothesis proportion test. We will therefore create a macro called *phyp2*. This macro assumes that the sample size is large.

Phyp2 Macro

```
gmacro
noecho
#
#   This macro is used in chapter 11
note calculates a z score for a given alpha for a sample proportion
note What is the hypothesized proportion?
set 'terminal' c50;
nobs=1.
end
note What is the sample proportion?
set 'terminal' c51;
nobs=1.
end
note What is the sample size of the data set?
set 'terminal' c52;
nobs=1.
end
note What is the alpha level?
set 'terminal' c53;
nobs=1.
end
let k50=c50
#k50 has the hypothesized portion number
let k51=c51
#k51 has the  sample portion
let k52=c52
#k52 has the sample size of the data set
let k53=c53/2
#k53 has the alpha level of a two-tail test.
let k55=(1-k53)
invcdf k55 k56;
normal 0 1.
let c60=((k51-k50)/(sqrt((k50*(1-k50))/k52)))*-1
let c61=(k51-k50)/(sqrt((k50*(1-k50))/k52))
note The lower and upper calculated z value for the sample
print c61 c60
let c65=k56
let c66=-k56
note These are the z values for the given alpha
print c66 c65
erase c60-c70
endmacro
```

CHAPTER 11

One thing to note is that in the *phy2* macro

```
let c60=((k51-k50)/(sqrt((k50*(1-k50))/k52)))*-1
let c61=(k51-k50)/(sqrt((k50*(1-k50))/k52))
```

Calculate the lower and upper test statistics for the population proportion.

Now we will use the *phyp2* macro to analyze the Francis Company data.

```
MTB > %phy2.mac
Executing from file: phy2.mac
calculates a z score for a given alpha for a sample proportion
What is the hypothesized proportion?
DATA> .80
What is the sample proportion?
DATA> .75
What is the sample size of the data set?
DATA> 200
What is the alpha level?
DATA> .05
The lower and upper calculated z value for the sample
```

Data Display

Row	C61	C60
1	-1.76777	1.76777

These are the z values for the given alpha

Data Display

Row	C66	C65
1	-1.95996	1.95996

Since the sample proportion is less than the hypothesized proportion, the test statistic will be negative. Because it will be negative, we will use the lower calculated Z value. We cannot reject the null hypothesis because the lower Z value for the sample is greater than the Z value for the given alpha.

This is illustrated below.

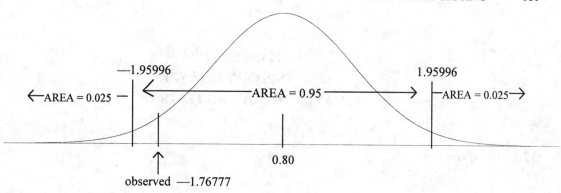

11.7 Statistical Summary

In this chapter we made inferences about a population parameter by checking if there is enough evidence to reject the null hypothesis. This is the same concept as in a court of law where a jury is supposed to convict someone only if there is enough evidence beyond a reasonable doubt. Neither the statistical method nor the court of law method is perfect. Both methods can make two types of mistakes in testing the null hypothesis. A Type I mistake is rejecting the null hypothesis when in fact it is true. A Type II mistake is accepting the null hypothesis when in fact it is false.

CHAPTER 12
ANALYSIS OF VARIANCE
AND CHI-SQUARE TESTS

12.1 Introduction

Often in statistics we are interested in whether two or more samples have the same mean from the same population. Statistics uses the concept of analysis of variance to answer this question.

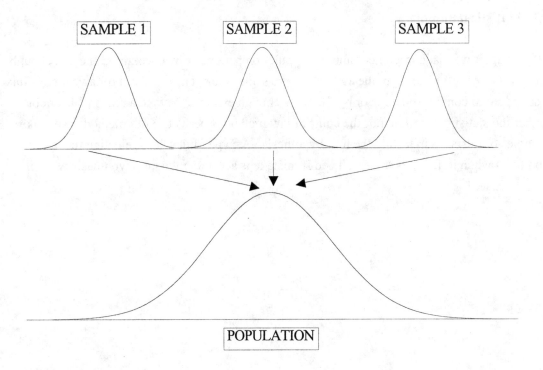

The first thing we must find out is whether the group of samples comes from the same population as shown in the above diagram or whether the group of samples does not come from the same population, as shown below.

ANALYSIS OF VARIANCE AND CHI-SQUARE TESTS   183

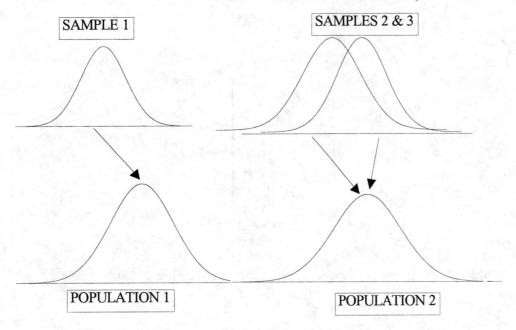

Analysis of variance, which is based on the concept of variation among samples, can be separated into two components, between-groups variation and within-group variation. This is shown below.

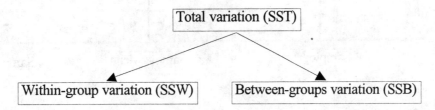

Total variation is the variation of all the samples about the overall mean of the samples.

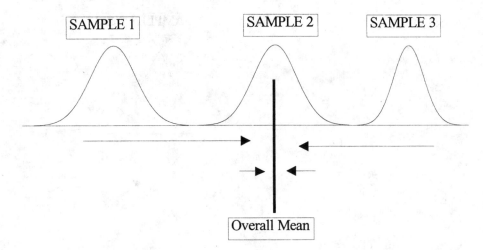

Within-group variation is variation about the mean of each individual sample.

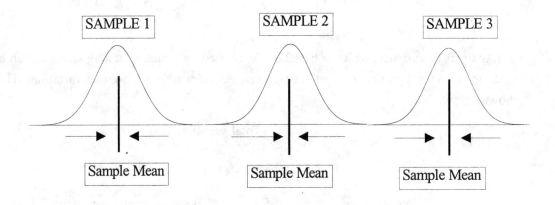

Between-groups variation is variation of group means about the overall mean.

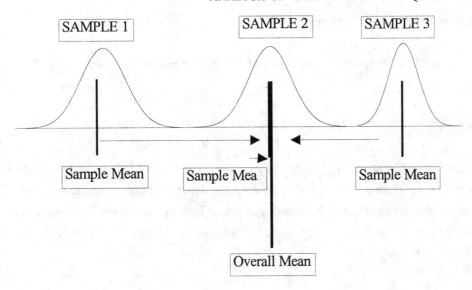

If the between-groups variation is significantly larger than the within-group variation, we can conclude that all the samples did not come from the same population. If the between-groups variation is not significantly larger, we cannot conclude that all the samples did not come from the same population.

Therefore, based on the above concepts the test statistic for the analysis of variance is

$$\frac{\text{between-groups variation}}{\text{within-groups variation}} = \frac{\text{between-groups mean square}}{\text{within-groups mean square}}$$

The above ratio follows an F distribution.

In the analysis of variance the general null and alternative hypotheses are

H_0: all the samples have the same mean
H_1: not all the samples have equal means.

12.2 One-Way Analysis of Variance

One-way analysis of variance studies the mean of two or more samples based on one classification.

Example 12.2A

A research institution says that gasoline is gasoline. There is no difference among different brands of gasoline in terms of mileage per gallon. An independent consumer rights organization did an experiment on three different brands of gasoline. Twenty cars of the same make and condition were divided into three groups, and each group was filled with a different brand of gasoline. The test results show:

	Gasoline	
A	B	C
34	29	32
28	32	34
29	31	30
37	43	42
42	31	32
27	29	33
29	28	

Is there any difference among the three brands of gasoline? Is there any significant difference at the 5% alpha level?

We will use the analysis of variance to see if there are any significant differences among the means. The null hypothesis and the alternative hypothesis would be

H_0: gasoline mean A = gasoline mean B = gasoline mean C
H_1: not all three gasoline means are equal

To do this analysis, let us put the data for gasoline A in column c1, gasoline B in column c2, and gasoline C in column c3.

When finished the worksheet should look like the following.

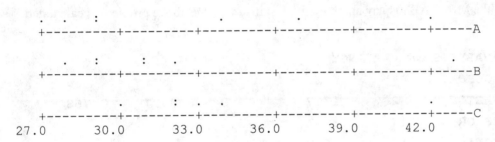

Let us first have a look at the dotplot of the three gasolines to get an overall picture of the data.

```
MTB > dotplot c1-c3;
SUBC> start 27 43.
```

Dotplot

```
         .  .     :                .              .                                .
        +---------+---------+---------+---------+---------+-----A
            .     :      :      .                                          .
        +---------+---------+---------+---------+---------+-----B
              .        :   .                                      .
        +---------+---------+---------+---------+---------+-----C
       27.0      30.0      33.0      36.0      39.0      42.0
```

From the dotplots we can see that the three gasolines get similar mileage. This suggests to us that the three gasoline brands might even have the same mean. We will test this by doing a one-way analysis of variance test. We will use the *aovoneway* MINITAB command to calculate the one-way analysis of variance.

MTB > aovoneway c1-c3

One-way Analysis of Variance

```
Analysis of Variance
Source     DF        SS        MS        F         P
Factor      2      13.7       6.8     0.27     0.768
Error      17     433.1      25.5
Total      19     446.8
                                      Individual 95% CIs For Mean
                                      Based on Pooled StDev
Level      N      Mean     StDev  --------+---------+---------+--------
A          7    32.286     5.589     (-------------*-------------)
B          7    31.857     5.113   (-------------*-------------)
C          6    33.833     4.215         (--------------*--------------)
                                      --------+---------+---------+--------
Pooled StDev =    5.048               30.0      33.0      36.0
```

From the MINITAB output we can see that the p-value for the one-way analysis is 0.768. Since this p-value is greater than an alpha value of 0.05, we cannot reject the null hypothesis that the three brands of gasoline have the same mean.

As stated above, the F statistic is defined as the between-groups variation divided by the within-group variation. Below we see where the between-groups and the within-group numbers are in the MINITAB display and how the F statistic for the above problem is calculated.

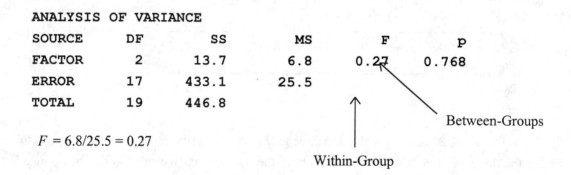

$F = 6.8/25.5 = 0.27$

We will now do the One-Way analysis of Variance in Excel.

Enter the data in EXCEL as shown below.

ANALYSIS OF VARIANCE AND CHI-SQUARE TESTS

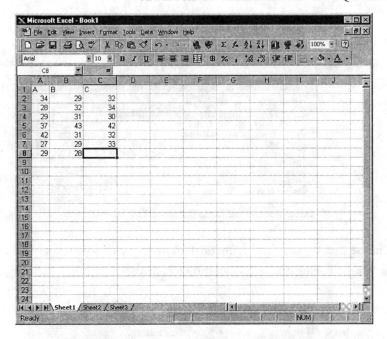

To do the one-way analysis of variance we will use the *Analysis Tools Pack* add-in.

The analysis tools pack will add *Data Analysis* menu item under the *Tools* menu as shown below.

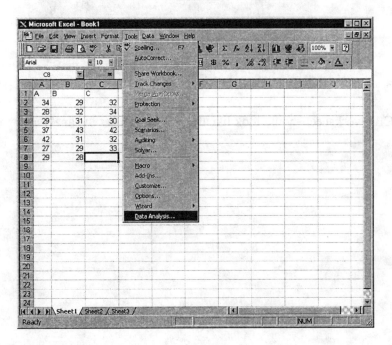

The *Data Analysis* menu item brings up the following dialog box.

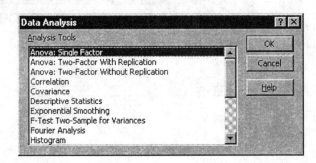

We will choose the *Anova: Single Factor* item to do the one-way analysis of variance. Fill out the Anova: Single Factor dialog box as shown below.

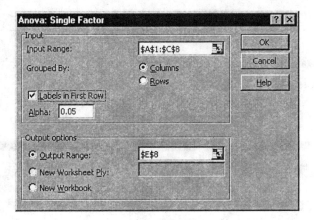

The EXCEL worksheet should look like the following after the analysis.

```
Anova: Single Factor

SUMMARY
  Groups       Count    Sum    Average       Variance
  A              7      226   32.28571429   31.23809524
  B              7      223   31.85714286   26.14285714
  C              6      203   33.83333333   17.76666667


ANOVA
  Source of Variation      SS          df      MS          F            P-value        F crit
  Between Groups       13.68095238     2    6.84047619   0.268489913   0.767714974   3.591537734
  Within Groups        433.1190476    17    25.47759104
  
  Total                446.8          19
```

Example 12.2B

There are many books to help people learn computer software packages. An instructor checked these books and found that they are all of similar quality. He picked four books and used a different book in each of his four classes. If the students all have the same average grades, he will use the cheapest book. The test results for students using the four books are:

	Classes		
1	2	3	4
43	77	72	72
45	72	73	74
67	75	71	75
68	69	65	65
73	67	68	66
72	66	69	68
55	65	73	74
62	63	72	81

Test to see if the four classes have the same grades at the 5% alpha level.

We will put class 1 in column c1, class 2 in column c2, class 3 in column c3, and class 4 in column c4. The worksheet should look like the following after entering the data.

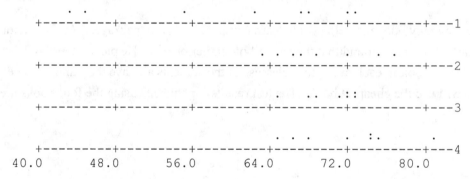

Let us look at a dotplot of the four classes.

```
MTB > dotplot c1-c4;
SUBC> start 41 81.
```

Dotplot

```
              . .            .             .           . .           . .
       +---------+---------+---------+---------+---------+-----1
                                      . . . .          .     . . .
       +---------+---------+---------+---------+---------+-----2
                                       .    . .    . :  :
       +---------+---------+---------+---------+---------+-----3
                                      . . .       .  : .       .
       +---------+---------+---------+---------+---------+-----4
      40.0      48.0      56.0      64.0      72.0      80.0
```

From the dotplot of the four classes we can see that the plots are somewhat similar. This suggests to us that the four classes might have the same average grade. We will test this by doing a one-way analysis of variance calculation.

Our null and alternative hypotheses would be

H_0: class 1 grade average = class 2 grade average = class 3 grade average = class 4 grade average
H_1: not all class averages are equal

```
MTB > aovoneway c1-c4
```

One-way Analysis of Variance

```
Analysis of Variance
Source      DF         SS         MS         F         P
Factor       3      612.8      204.3      4.10     0.016
Error       28     1394.1       49.8
Total       31     2007.0
                                          Individual 95% CIs For Mean
                                          Based on Pooled StDev
Level       N       Mean      StDev    --------+---------+---------+--------
1           8     60.625     11.747    (--------*--------)
2           8     69.250      4.979                      (--------*--------)
3           8     70.375      2.825                         (--------*--------)
4           8     71.875      5.330                             (--------*-------)
                                       --------+---------+---------+--------
Pooled StDev =     7.056                    60.0      66.0      72.0
```

From the MINITAB output we can see that the calculated p-value is 0.016. Therefore, since the p-value is less than the alpha value of 0.05, we reject the null hypothesis that every class has the same average grade.

The MINITAB output also shows the calculated mean grade for each class and the 95% confidence interval of mean grade for each class. From the diagram we conclude that the class with a statistically different mean grade is class 1. This conclusion stems from a graphical approach.

There are also numerical approaches. The advantage of numerical approaches is that they are more precise. We will now look at numerical approaches to test which class has statistically different grades.

To do this we will have to use the *oneway* MINITAB command. For the *oneway* command we are going to have to put all the data in column c1 and indicate the different classes in column c2.

After entering the class data, the worksheet should look like the following.

Notice that every data item in column c1 is assigned to a specific class in column c2.

We use the Tukey method to determine which samples are statistically different from the other samples. For this method, MINITAB uses the *tukey* subcommand.

```
MTB > oneway c1 c2;
SUBC> tukey.
```

One-way Analysis of Variance

```
Analysis of Variance for C1
Source     DF         SS         MS         F         P
C2          3      612.8      204.3      4.10     0.016
Error      28     1394.1       49.8
Total      31     2007.0

                                    Individual 95% CIs For Mean
                                    Based on Pooled StDev
Level       N       Mean      StDev  --------+---------+---------+--------
1           8     60.625     11.747  (-------*--------)
2           8     69.250      4.979                   (-------*--------)
3           8     70.375      2.825                     (-------*--------)
4           8     71.875      5.330                       (--------*-------)
                                     --------+---------+---------+--------
Pooled StDev =     7.056              60.0      66.0      72.0

Tukey's pairwise comparisons

    Family error rate = 0.0500
Individual error rate = 0.0108
```

```
Critical value = 3.86

Intervals for (column level mean) - (row level mean)
                    1             2             3
        2       -18.255
                  1.005

        3       -19.380       -10.755
                 -0.120         8.505

        4       -20.880       -12.255       -11.130
                 -1.620         7.005         8.130
```

The Tukey method calculates the confidence interval between two groups. To determine which two groups are statistically different we see if the interval contains zero or not. If it does, then it is likely that the two groups are not significantly different. For example, the 95% confidence interval for groups 2 and 3 is

$$-10.755 < \mu_2 - \mu_3 < 8.505$$

Since this interval contains zero, we conclude that we cannot reject that class 2 and class 3 are statistically different.

From the Tukey matrix we can also see that class 1 and class 3 are statistically different because their interval does not contain 0. Also class 1 and class 4 are statistically different because their interval does not contain 0.

This illustrates that we might not get the same answer using both a numerical method and a graphical method. Graphically we saw that class 1 was statistically different from every other class, but numerically we saw that it was not statistically different from class 2.

So far we have looked at examples based on a single classification. The first example analyzed the mean based on gasoline. The second example analyzed the mean based on classes. These classifications are called *factors*. In the gasoline example, we had three different types of gasoline. In the classroom example, we had four classes. The individual types of gasoline and types of class are called *levels*.

12.3 Two-Way Analysis of Variance

Doing an analysis of variance with one factor is called a one-way analysis of variance. We will now expand the analysis of variance and test to see if the means from different samples are the same or not based on two factors. Analysis of variance of two factors is called a two-way analysis of variance.

We can think of factors as influencing the value of the mean of each sample. For example, in the gasoline example, we were interested in miles per gallon. The factor that we thought might be influencing the mileage per gallon was the different brands. Now we are interested in two-factor analysis, in two things affecting the mean that we are interested in.

Example 12.3A

Suppose we are interested to know if years of work experience and geographical location influence annual salary. The data, given below, show salaries in thousands of dollars.

	Years of Work Experience		
Region	1	2	3
1. West	16, 16.5	19, 17	24, 25
2. Midwest	21, 20.5	20, 19	21, 22.5
3. Northeast	18, 19	21, 20.9	22, 21
4. South	13, 13.5	20, 20.8	25, 23

Here we are interested in the mean of the annual salary. We believe that two factors influence the mean: years of work experience and region of the country. It turns out that we should also look at the influence of the interaction of these two factors.

The variability of annual salary can thus be broken down in the following way:

Total Variation = Variation of Factor A + Variation of Factor B + Variation of Interaction of Factors AB + Random Error

In a two-way analysis of variance, there are three tests in which we are interested. In our salary example they would be the following:

1. To test the hypothesis of no difference due to years of working experience:

H_0: mean year 1 = mean year 2 = mean year 3
H_1: not all mean years are equal

The test would follow an F distribution and would have the following ratio:

$$F = \frac{\text{variation of factor A (years of working experience)}}{\text{random error}}$$

2. To test the hypothesis of no difference due to geographic region:

H_0: mean region 1 = mean region 2 = mean region 3
H_1: not all mean regions are equal

The test would follow an F distribution and would have the following ratio:

$$F = \frac{\text{variation of factor B (geographic region)}}{\text{random error}}$$

3. To test the hypothesis of no difference due to interaction of work experience and region:

H_0: interaction of work experience and geographic region has no effect
H_1: interaction of work experience and geographic region has an effect

The test would follow an F distribution and would have the following ratio:

$$F = \frac{\text{variation of interaction of factor A and factor B (experience and region)}}{\text{random error}}$$

We will now test our annual salary example using the *twoway* command. The worksheet should look like the following after the data are entered.

	C1	C2	C3	C4
1	16.0	1	1	
2	16.5	1	1	
3	19.0	1	2	
4	17.0	1	2	
5	24.0	1	3	
6	25.0	1	3	
7	21.0	2	1	
8	20.5	2	1	
9	20.0	2	2	
10	19.0	2	2	
11	21.0	2	3	
12	22.5	2	3	
13	18.0	3	1	
14	19.0	3	1	

	C1	C2	C3	C4
14	19.0	3	1	
15	21.0	3	2	
16	20.9	3	2	
17	22.0	3	3	
18	21.0	3	3	
19	13.0	4	1	
20	13.5	4	1	
21	20.0	4	2	
22	20.8	4	2	
23	25.0	4	3	
24	23.0	4	3	
25				
26				
27				

```
MTB > twoway c1 c2 c3;
SUBC> mean c2 c3.
```

Two-way Analysis of Variance

```
Analysis of Variance for salary
Source          DF        SS         MS        F         P
region           3     7.921      2.640     4.05     0.033
work             2   132.903     66.452   101.91     0.000
Interaction      6    77.730     12.955    19.87     0.000
Error           12     7.825      0.652
Total           23   226.380

                          Individual 95% CI
region        Mean      ------+---------+---------+---------+-----
1            19.58            (---------*---------)
2            20.67                              (---------*---------)
3            20.32                     (---------*---------)
4            19.22      (----------*---------)
                        ------+---------+---------+---------+-----
                          18.90     19.60     20.30     21.00

                          Individual 95% CI
work          Mean      --------+---------+---------+---------+---
1            17.19      (--*--)
2            19.71                (---*--)
3            22.94                                   (--*--)
                        --------+---------+---------+---------+---
                            18.00     20.00     22.00     24.00
```

```
MTB > let k1=2.640/.652
MTB > print k1
K1         4.04908
MTB > invcdf .95;
SUBC> f 3 12.
    0.9500     3.4903
MTB > let k2=66.452/.652
MTB > print k2
K2         101.920
MTB > invcdf .95;
SUBC> f 2 12.
    0.9500     3.8853
MTB > let k3=12.955/.652
MTB > print k3
K3         19.8696
MTB > invcdf .95;
SUBC> f 6 12.
    0.9500     2.9961
```

It is unfortunate that the *twoway* command does not calculate any F statistics or p-values. Therefore we are going to have to calculate our own F statistics.

Let us first test to see if work region has no effect on annual salary. The calculated F value would be

$$\text{region/error} = 2.640/0.652 = 4.04908$$

This is done by the MINITAB expression, **let k1=2.640/.652**. The *print* command prints the value of the constant, k1, that has the F value. The critical value of an F distribution with 3,12 degrees of freedom at 0.05 value is 3.4903. Because the calculated value of 4.04908 is greater than the critical value of 3.4903, we can conclude that the geographic region does affect annual salary.

The region variance, 2.640, and the error variance, 0.652, were obtained from the MINITAB output in the analysis of variance section. The region degree of freedom, 3, and the error degree of freedom, 12, were also obtained from the MINITAB output in the analysis of variance section.

```
MTB > let k2=66.452/.652
MTB > print k2
K2         101.920
MTB > invcdf .95;
SUBC> f 2 12.
    0.9500     3.8853
```

Next we test to see if the work experience affects annual salary. The calculated F value would be work/error = 66.452/0.652 = 101.920

This is done by the MINITAB expression, **let k2=66.452/.652**. The *print* command prints the value of the constant, k2, that has the *F* value. The critical value of an *F* distribution with 2,12 degrees of freedom at 0.05 value is 3.8853. Because the calculated value of 101.920 is greater than the critical value of 3.8853, we conclude that work experience does affect annual salary.

```
MTB > let k3=12.955/.652
MTB > print k3
K3       19.8696
MTB > invcdf .95;
SUBC> f 6 12.
   0.9500    2.9961
```

Next we test to see if the interaction of work experience and geographic region affects annual salary. The calculated *F* value would be

$$\text{work/error} = 12.955/0.652 = 19.8696$$

This is done by the MINITAB expression, **let k3=12.955/.652**. The *print* command prints the value of the constant, k3, that has the *F* value. The critical value of an *F* distribution with 6, 12 degrees of freedom at 0.05 value is 2.9961. Because the calculated value is greater than the critical value of 2.9961, we conclude that the interaction of work experience and geographic regions affects annual salary.

12.4 Chi-Square Test

Example 12.4A

We are now interested in checking to see whether two variables are independent of each other or not.

Suppose someone has asked us if majors and grades are related. The data for this question are shown below:

	Grades			
Major	A	B	C	F
Science	12	36	34	8
Humanities	10	24	46	10
Business	8	30	70	12

The hypotheses would be

H_0: major and grades are independent
H_1: major and grades are not independent

For this problem, we would follow a chi-square test:

```
MTB > chisquare c1-c4
```

Chi-Square Test

```
Expected counts are printed below observed counts

                A         B         C         F      Total
    1          12        36        34         8         90
             9.00     27.00     45.00      9.00

    2          10        24        46        10         90
             9.00     27.00     45.00      9.00

    3           8        30        70        12        120
            12.00     36.00     60.00     12.00

Total          30        90       150        30        300

Chi-Sq =   1.000 +   3.000 +   2.689 +   0.111 +
           0.111 +   0.333 +   0.022 +   0.111 +
           1.333 +   1.000 +   1.667 +   0.000 = 11.378
DF = 6,  P-Value = 0.077
```

```
MTB > invcdf .95;
SUBC> chisquare 6.
```

Inverse Cumulative Distribution Function

```
Chi-Square with 6 DF

 P( X <= x)          x
    0.9500       12.5916
```

We conclude that grades and major are independent, because the calculated chi-square statistic of 11.378 is less than the critical value of 0.05 alpha, which is 12.5916.

12.5 Statistical Summary

In this chapter we use either one-way or two-way ANOVA to see if several sample means are statistically different from each other. In addition, we discussed how the chi-square statistic can be used to test whether two variables are independent of each other.

CHAPTER 13
SIMPLE LINEAR REGRESSION
AND THE CORRELATION COEFFICIENT

13.1 Introduction

In previous chapters we have been primarily interested in analyzing single variables. In this chapter we are interested in analyzing two variables and the relationship between them. The two techniques that we will look at are regression analysis and correlation analysis.

13.2 Regression Analysis

In regression analysis, it is believed that one variable can explain for another variable in a linear fashion. Mathematically this equation looks like the following:

$$y = a + bx$$

The two variables of interest in this equation are x and y. In this equation x is the independent variable and y is the dependent variable.

In regression analysis we are modeling the relationship between two variables and then using the model to predict the value of the dependent variable from the value of the independent variable.

Graphically the relationship between the two variables is believed to look like the following:

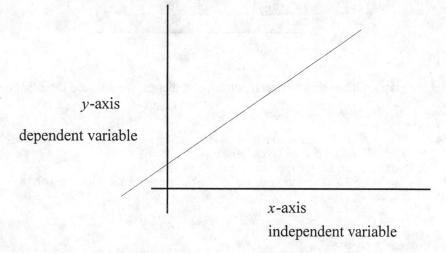

Example 13.2A

Let us now look at a problem about the relationship between height and weight. The height and weight of six children are given below.

Children's Weight and Height	
y, Pound	x, Inches
92	55
95	56
99	57
97	58
102	59
104	60

Let us use a MINITAB graph to see if the two variables, pounds and inches, have a linear relationship. To do this, we must first put the y data in column c1 and the x data in column c2, and we will name column c1 '*pounds*' and column c2 '*inches*'. The worksheet should look like the following when we finish entering the data.

	C1	C2	C3
	pounds	inches	
1	92	55	
2	95	56	
3	99	57	
4	97	58	
5	102	59	
6	104	60	
7			
8			

To get the graph relationship between pounds and inches, we will use the *plot* command.

```
MTB > gstd

* NOTE  * Standard Graphics are enabled.
          Professional Graphics are disabled.
          Use the GPRO command to enable Professional Graphics.
MTB > plot c1 c2.
```

```
Plot

    104.0+                                                    *
 pounds -
        -                                         *
        -
        -
    100.0+                             *
        -
        -                                  *
        -
        -
     96.0+              *
        -
        -
        -
     92.0+    *
        -
          ----+---------+---------+---------+---------+---------+--inches
            55.0      56.0      57.0      58.0      59.0      60.0
```

In the above graph we can see that the relationship is not perfectly linear. This means that inches do not explain 100% of the variation in pounds. Let's do a regression analysis of the extent to which inches explain the variation in pounds. To do this, we are interested in finding a line that best fits the above graph. We will use a method called *least squares*. What least squares does is to calculate the *a* and *b* of a linear equation that best fits the data. The formula to calculate *a* and *b* is shown below

$$a = \bar{y} - b\bar{x} = \frac{\sum_{i=1}^{n} y - b(\sum_{i=1}^{n} x_i)}{n}$$

$$b = \frac{\sum_{i=1}^{n}(x_i - \bar{x})(y_i - \bar{y})}{\sum_{i=1}^{n}(x_i - \bar{x})^2} = \frac{s_{xy}}{s_x^2}$$

In MINITAB, we use the *regress* command to calculate the *a* and *b* of a linear line.

```
MTB > regress c1 1 c2
```

Regression Analysis

```
The regression equation is
pounds = - 31.6 + 2.26 inches

Predictor        Coef       StDev          T        P
Constant       -31.62       21.39      -1.48    0.213
inches         2.2571      0.3718       6.07    0.004

S = 1.555      R-Sq = 90.2%     R-Sq(adj) = 87.8%

Analysis of Variance

Source            DF         SS          MS         F        P
Regression         1     89.157      89.157     36.86    0.004
Residual Error     4      9.676       2.419
Total              5     98.833
```

The MINITAB output indicates that the best fitting linear equation is

```
pounds = - 31.6 + 2.26 inches.
```

This equation states that for every 1-inch increase in height there is a 2.26-pound increase in weight.

Let us now do the regression in EXCEL. Enter the data in EXCEL as shown below,

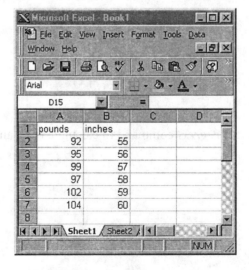

To do the regression choose **Tools-> Data Analysis** to get the *Data Analysis* dialog box. In the *Data Analysis* dialog box, choose the *regression* option as shown below,

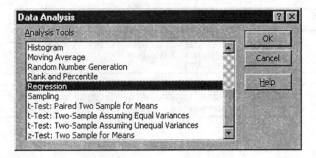

Then fill out the *Regression* dialog box as shown below. When finished, pressed the *OK* button.

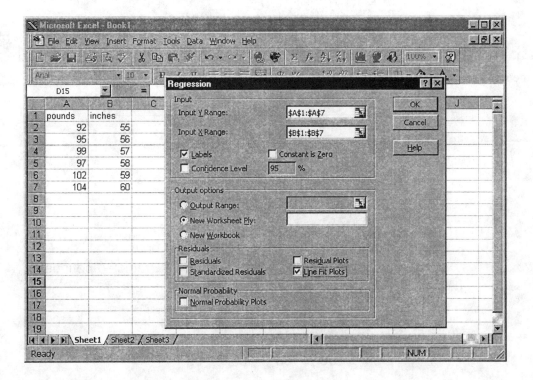

EXCEL produces the following regression report.

SUMMARY OUTPUT

Regression Statistics	
Multiple R	0.949787282
R Square	0.902095881
Adjusted R Square	0.877619851
Standard Error	1.555328782
Observations	6

ANOVA

	df	SS	MS	F	Significance F
Regression	1	89.15714286	89.15714286	36.85629921	0.003718675
Residual	4	9.676190476	2.419047619		
Total	5	98.83333333			

	Coefficients	Standard Error	t Stat	P-value	Lower 95%	Upper 95%	Lower 95.0%	Upper 95.0%
Intercept	-31.61904762	21.38762225	-1.478380684	0.213382586	-91.00072972	27.76263448	-91.00072972	27.76263448
inches	2.257142857	0.371794691	6.070938907	0.003718675	1.224873168	3.289412546	1.224873168	3.289412546

RESIDUAL OUTPUT

Observation	Predicted pounds	Residuals
1	92.52380952	-0.523809524
2	94.78095238	0.219047619
3	97.03809524	1.961904762
4	99.2952381	-2.295238095
5	101.552381	0.447619048
6	103.8095238	0.19047619

The following chart was also produced, because we chose the *Line Fit Plots* option in the *Regression* dialog box.

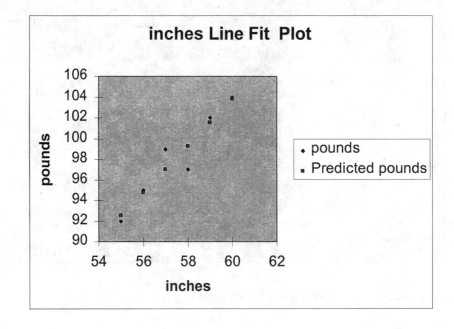

SIMPLE LINEAR REGRESSION AND THE CORRELATION COEFFICIENT

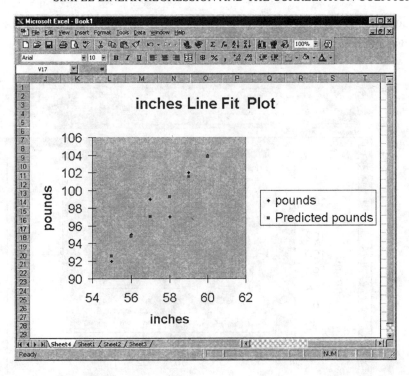

EXCEL produces the chart with a gray background. Let us change the background to white. To do this move the move cursor over the gray background as shown below.

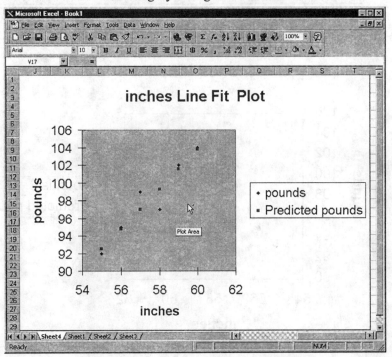

210   CHAPTER 13

Then double click the left mouse button to get the *Format Plot Area* dialog box. In the *Area* section choose the *None* option as shown below

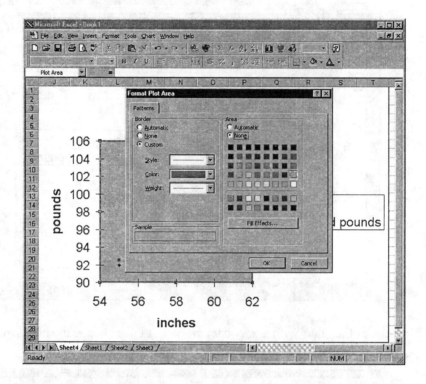

The resulting chart is shown below

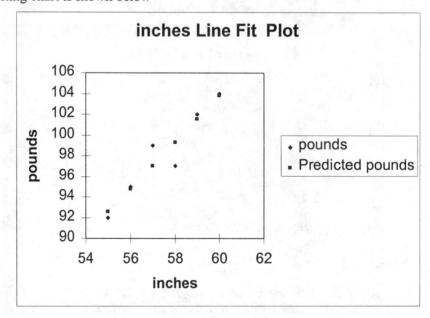

13.3 Coefficient of Determination

We saw in the plot of the weight/height data that the relationship between them is not perfectly linear. This means that there will not be an exact 2.26-pound increase for every 1-inch increase in height. If this did happen, it would mean that inches explain 100% of the variation in pounds. MINITAB output can give us the extent to which inches explain the variation in pounds. In this case we see that inches explain the variation in pounds 90.2% of the time. The section of the output that gives this information is reproduced below.

```
s = 1.555        R-sq = 90.2%      R-sq(adj) = 87.8%
```

The 90.2% is called the R-square of a regressed linear equation. It is also called the coefficient of determination.

The 90.2% can be calculated using the information from the MINITAB output. The part necessary for calculating the R-square is reproduced below.

Analysis of Variance

```
SOURCE          DF           SS           MS           F          p
Regression      1        89.157       89.157       36.86      0.004
Error           4         9.676        2.419
Total           5        98.833
```

The total variation of the dependent data is represented by the number 98.833. This number is obtained from the intersection of the SS column and the Total row. The regression line variation is represented by the number 89.157. This number is obtained from the intersection of the SS column and the Regression row. The R-square number is obtained from these two numbers. It is calculated as follows

$$R\text{-}square = \frac{89.157}{98.833} = 0.902$$

The MINITAB output also indicates how much of the total variation of the data is not explained by the regressed equation. This is represented by the number 9.676. This number is obtained from the intersection of the SS column and the Error row. Note that

Total = Regression + Error = 98.833 = 89.157 + 9.676

212   CHAPTER 13

Many books like to give the above equation as follows:

SST   =   SSE   +   SSR
Total Variation   Unexplained Variation   Explained Variation

R-square can be represented graphically by a Venn diagram. This is done by one circle representing the total variation and another circle representing the regression variation. The portion of the total circle that is intersected by the regressed circle represents the R-square. The portion of the total circle that is not part of the regression circle represents the part of the variation that is not explained by the independent variable. This portion is also called the *error* of the regressed line. The Venn diagram below illustrates the height/weight data.

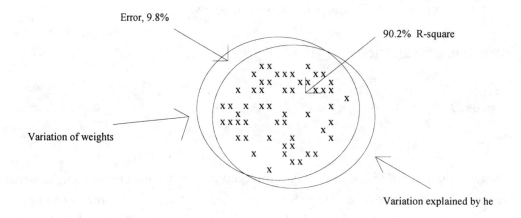

13.4 Correlation Coefficient

Correlation analysis is another way to study the relationship between two variables. In particular, it measures the strength of association between two variables. Correlation is expressed in values between –1 and +1. When $r = +1$, then x and y exist in a perfectly positive linear relationship; $r = -1$ means that the two variables x and y exist in a perfectly negative linear relationship; and $r = 0$ means that x and y are not linearly related. These concepts are graphically illustrated below.

SIMPLE LINEAR REGRESSION AND THE CORRELATION COEFFICIENT 213

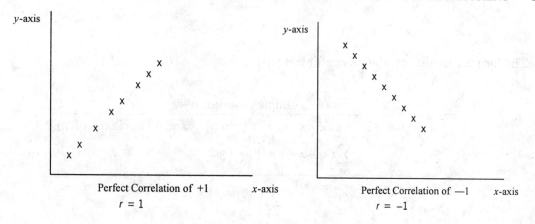

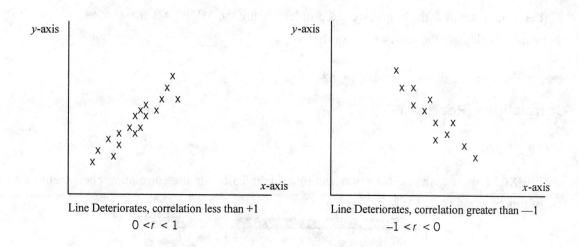

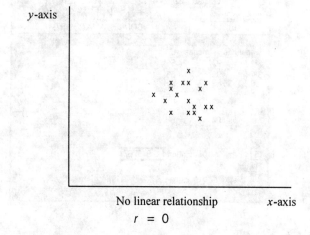

The formula for the correlation coefficient is

$$r = \frac{\text{sample covariance}}{(\text{sample standard deviation of } x)(\text{sample standard deviation of } y)}$$

$$= \frac{\frac{1}{n-1}\sum_{i=1}^{n}(x_i - \bar{x})(y_i - \bar{y})}{[\frac{1}{n-1}\sum_{i=1}^{n}(x_i - \bar{x})^2]^{1/2}[\frac{1}{n-1}\sum_{i=1}^{n}(y_i - \bar{y})^2]^{1/2}} = \frac{s_{xy}}{s_x s_y}$$

Let us look again at the height/weight problem, using the MINITAB *correlation* command to calculate the correlation coefficient.

```
MTB > correlation c1 c2
```

Correlations (Pearson)

```
Correlation of pounds and inches = 0.950, P-Value = 0.004
```

In EXCEL there is the *correl* worksheet function to calculate the correlation coefficient. The use of this function is shown below.

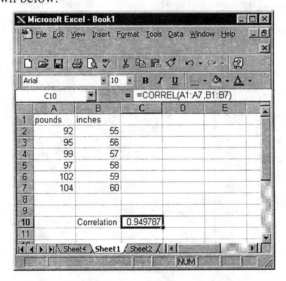

From the correlation coefficient value, 0.95, we can see that there is a strong positive relationship between pounds and inches.

If we look carefully at the graphical representation of regression analysis and correlation analysis, we might find that they are very similar. Because of these similarities, we might ask if there is any relationship between regression analysis and correlation analysis. In fact, there is a relationship, as expressed by the following equation

$$\text{R-square} = r^2$$

We can check this relationship by squaring our calculated correlation coefficient of 0.95. This would be $0.95 * 0.95 = 0.902$, which equals our R-square.

13.5 REGRESSION EXAMPLES

Example 13.5A

The Organization of Petroleum Exporting Countries (OPEC) has tried to control the price of crude oil since 1973. The price of crude oil rose dramatically from the mid-1970s to the mid-1980s. As a result, motorists were confronted with a similar upward spiral of gasoline prices. The following table presents the average unit prices of gasoline and crude oil from 1975 to 1988.

Price of Gasoline and Crude Oil		
Year, i	Gasoline, y cents/gallon	Crude Oil, x $/barrel
1975	57	7.67
1976	59	8.19
1977	62	8.57
1978	63	9.00
1979	86	12.64
1980	119	21.59
1981	133	31.77
1982	122	28.52
1983	116	26.19
1984	113	25.88
1985	112	24.09
1986	86	12.51
1987	90	15.40
1988	90	12.57

216   CHAPTER 13

Use MINITAB to do a regression analysis of the relationship between the price of a gallon of gasoline and the price of a barrel of crude oil.

We will put the gasoline data in column c1 and the crude oil data in column c2. We will name column c1 '*gasoline*' and column c2 '*oil*'.

The first thing that we will do in trying to understand the relationship between the price of gasoline and the price of a barrel of oil is to plot the two variables. As shown below, using the *plot* command, we can see that the relationship is positively linear.

```
MTB > gstd
* NOTE  * Standard Graphics are enabled.
          Professional Graphics are disabled.
          Use the GPRO command to enable Professional Graphics.
MTB > plot c1 c2
```

Plot

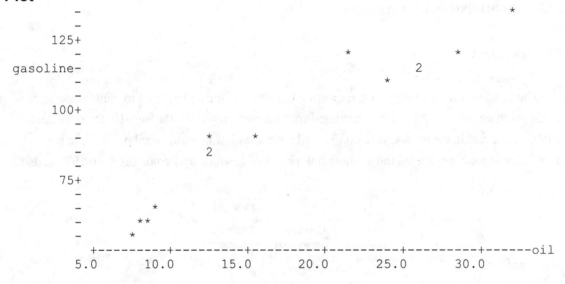

We will now regress the gasoline and oil variables.

```
MTB > regress c1 1 c2
```

Regression Analysis

```
The regression equation is
gasoline = 41.9 + 2.95 oil

Predictor         Coef        StDev           T         P
Constant        41.917        4.558        9.20     0.000
oil             2.9485       0.2361       12.49     0.000

S = 7.252       R-Sq = 92.9%     R-Sq(adj) = 92.3%

Analysis of Variance

Source              DF          SS          MS         F         P
Regression           1      8202.3      8202.3    155.95     0.000
Residual Error      12       631.2        52.6
Total               13      8833.4
```

The estimated regression equation according to the *regress* command is

$$\text{GASOLINE} = 41.9 + 2.95 \text{ OIL}$$

This regression equation states that for every dollar increase in the price of a barrel of crude oil, the price of a gallon of gasoline will increase 2.95 cents.

The R-square is 92.9%. This means that the price of crude oil accounts for 92.9% of the variability in the price of gasoline.

Example 13.5B

Suppose we have random samples of 10 households showing the numbers of cars per household, as shown below.

218   CHAPTER 13

Numbers of Cars per Household		
Household	Cars, y	People, x
1	4	6
2	1	2
3	3	4
4	2	3
5	2	4
6	3	4
7	4	6
8	1	3
9	2	2
10	2	2

Use MINITAB to do a regression analysis to investigate the relationship between the number of people in a household and the number of cars in a household.

We will put the car data in column c1 and the people data in column c2. We will name column c1 '*cars*' and column c2 '*people*'.

The first thing that we will do in trying to understand the relationship between the number of people in a household and the number of cars is to plot the two variables. As shown below we can see that the relationship is positively linear.

```
MTB > gstd
* NOTE * Standard Graphics are enabled.
         Professional Graphics are disabled.
         Use the GPRO command to enable Professional Graphics.
MTB > plot c1 c2
```

Plot

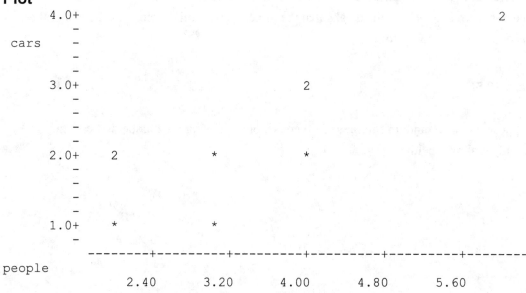

```
MTB > regress c1 1 c2
```

Regression Analysis

```
The regression equation is
cars = 0.176 + 0.618 people

Predictor        Coef       StDev          T        P
Constant       0.1765      0.4905       0.36    0.728
people         0.6176      0.1266       4.88    0.001

S = 0.5720      R-Sq = 74.8%     R-Sq(adj) = 71.7%
```

Analysis of Variance

```
Source          DF         SS          MS         F        P
Regression       1     7.7824      7.7824     23.78    0.001
Residual Error   8     2.6176      0.3272
Total            9    10.4000
```

The estimated regression equation according to the *regress* command is

```
cars = 0.176 + 0.618 people
```

This regression equation states that for every additional person in a household there will be a 0.618 increase in the number of cars.

The R-square is 74.8%. This means that the number of persons explains 74.8% of the variability in the number of cars in the household.

Notice that this R-square is less than the R-square of the gasoline problem. This result is confirmed by the fact that the plot in this problem is less linear than the plot in the gasoline problem.

13.6 Statistical Summary

In this chapter we look at two methods of analyzing the relationship between two variables. The two methods are the simple linear regression method and the correlation method. The simple linear regression analysis can be used to measure how much one variable is explained by another variable. The correlation analysis can be used to measure the strength of the linear relationship between two variables.

CHAPTER 14
SIMPLE LINEAR REGRESSION AND CORRELATION: ANALYSES AND APPLICATIONS

14.1 Introduction

When we do regression analysis, there are six underlying assumptions:

1. The dependent and independent variables have a linear relationship.
2. The expected value of the error term is zero.
3. The variance of the error term is constant.
4. The error terms are independent.
5. The independent variables are uncorrelated.
6. The errors are normally distributed.

In general, we use sample instead of population data to estimate regression coefficients. If the regression model follows the above six assumptions, then we can make inferences about the population regression coefficients from the sample regression coefficients. In the last chapter, we used regression and correlation analysis to relate two variables on the basis of sample information. But data from a sample represent only part of the total population. Because of this, we may think of our estimated sample regression line as an estimate of a true, but unknown population regression line of the form:

$$y = \alpha + \beta x$$

Therefore, what we are doing is using the sample regression line

$$y = a + bx$$

to estimate the population regression line.

14.2 Two-Tail *t* Test for β

One statistical inference that we are interested in is to test the hypothesis about the value of β. Specifically, we want to know whether β is equal to zero. If β does equal to zero, then this would mean that the independent variable actually does not explain any part of the variation of the dependent variable. The null hypothesis (H_0) and the alternative hypothesis (H_1) would be the following.

$$H_0: \beta = 0$$
$$H_1: \beta \neq 0$$

We will use the *t* distribution to test the above hypothesis. The test statistic is shown below.

$$t = \frac{\frac{\sum_{i=1}^{n}(x_i - \bar{x})(y_i - \bar{y})}{\sum_{i=1}^{n}(x_i - \bar{x})^2} - 0}{\frac{s_e}{\sqrt{\sum_{i=1}^{n}(x_i - \bar{x})^2}}} = \frac{b - 0}{s_b}$$

where s_e and s_b are the sample standard deviation of the error terms and standard error of the slope, respectively.

We will use the height/weight problem from Chapter 13 to study the inference of β. The *regress* command calculates the above *t* test statistic. Below is a reproduction of the *regress* command and its output that we did in Chapter 13.

```
MTB > regress c1 1 c2
```

Regression Analysis

```
The regression equation is
pounds = - 31.6 + 2.26 inches

Predictor         Coef        StDev            T         P
Constant        -31.62        21.39        -1.48     0.213
inches          2.2571       0.3718         6.07     0.004

S = 1.555      R-Sq = 90.2%      R-Sq(adj) = 87.8%

Analysis of Variance

Source            DF           SS           MS         F         P
Regression         1       89.157       89.157     36.86     0.004
Residual Error     4        9.676        2.419
Total              5       98.833
```

The *regress* command calculated the *t* test statistic as 6.07. Usually we would have to check a *t* table to find the associated *t* value for an alpha value of 0.05. Then we would test to see if the calculated sample *t* value was greater or less than the value in the *t* table.

With the advent of computer software, however, the decision process is much simpler. Most statistical software also calculates the *p*-value associated with the calculated *t* value. To reject or accept the null hypothesis, we only need to compare the calculated *p*-value with the alpha value that we are interested in.

If we want to test at the 5% alpha level, we can reject the null hypothesis that $\beta = 0$ because the *p*-value associated with the test statistic is 0.004, which is smaller than the 0.05 alpha value.

We will reject the null hypothesis any time the *p*-value is less than the alpha value.

14.3 Two-Tail t Test for α

We can also test whether the population intercept α is significantly different from zero or not. As with the test for the population slope, β, we can use the regression output to test the significance of α of a linear equation. We will use the t statistic defined on the previous page to test whether α is significantly different from zero or not. The null and alternative hypotheses for α would be:

$$H_0: \alpha = 0$$
$$H_1: \alpha \neq 0$$

The part of the regression output that is necessary for our analysis is

```
Predictor        Coef       Stdev     t-ratio         p
Constant       -31.62       21.39       -1.48     0.213
```

We can see that the calculated t value for α is -1.48 and the associated p-value is 0.213. If the alpha value that we are interested in is 0.05, we cannot reject the null hypothesis because the p-value is larger than the alpha value of 0.05.

14.4 Confidence Interval of β

Another way to make inferences about the population is to create a confidence interval. If we are interested in a 95% confidence interval, this would mean that 95% of the confidence intervals created will contain the true population parameter. The concept of confidence intervals was discussed in Chapter 10.

In MINITAB there is no command to create a confidence interval for β in a regressed equation, and so we must create our own macro. We will call our macro *bintreg*. The formula for the β confidence interval is

$$b - t_{\alpha/2} s_b \leq \beta \leq b + t_{\alpha/2} s_b$$

where β is the population slope and b is the sample slope.

Bintreg Macro

```
Gmacro
noecho
erase c50-c70
#this macro is used in chapter 14
#this interval is used to calculate the confidence
#interval for the coefficient of a regressed equation
note What is the s for the confidence interval?
set 'terminal' c50;
nobs=1.
end
let k50=c50
note What is the confidence (%) for the confidence interval?
set 'terminal' c51;
nobs=1.
end
let k51=c51
note What is the beta coefficient for the confidence interval?
set 'terminal' c52;
nobs=1.
end
let k52=c52
let k60=sum(('x'-mean('x'))**2)
let k61=k50/sqrt(k60)
let k62=n('x')-2
let k64=(1+k51)/2
invcdf k64 k65;
t k62.
let c71=k52+k65*k61
let c70=k52-k65*k61
note confidence interval is:
name c70 'lower' c71 'upper'
print c70 c71
erase c50-c90
endmacro
```

We will now illustrate the use of the *bintreg* macro on the height/weight problem discussed in Chapter 13. When we use the *bintreg* macro we need to name the independent variable as 'x'. If we do not, the *bintreg* macro will not work. The worksheet should look like the following:

	C1	C2
	pounds	x
1	92	55
2	95	56
3	99	57
4	97	58
5	102	59
6	104	60

We will find a 95% confidence interval for β using the macro *bintreg*.

```
MTB > %bintreg
Executing from file: bintreg.MAC
What is the s for the confidence interval?
DATA> 1.555
What is the confidence (%) for the confidence interval?
DATA> .95
What is the beta coefficient for the confidence interval?
DATA> 2.2571
confidence interval is:

Data Display

 Row    lower      upper

  1    1.22505    3.28915
```

From the *bintreg* macro we can see that the 95% confidence interval is from 1.22505 pounds to 3.28915 pounds.

14.5 *F* Test

We used the *t* test to see if β is significantly different from zero. We can also use the *F* test to see if β is significantly different from zero. Below is a reproduction of the part of the regression output that we will need to use the *F* test.

Analysis of Variance

```
Source           DF        SS         MS        F        P
Regression        1     89.157     89.157    36.86    0.004
Residual Error    4      9.676      2.419
Total             5     98.833
```

From the regression output we can see that the calculated F value was 36.86 for the height/weight problem and the associated p-value was 0.004. If we use an alpha value of 0.05 we can conclude that β is significantly different from 0 and that the height does indeed help explain the variation in weight.

14.6 The Relationship between the *F* Test and the *t* Test

One question that we might ask is, 'If we are using both the F test and the t test to test the same hypothesis, is there any relationship between the F test and the t test?' The answer is yes. For a simple regression, the mathematical relationship is

$$F = t^2$$

Our t value is 6.07 and our F value is 36.86. If we square our t value we get

$$t^2 = 6.07^2 = 36.8449$$

The slight discrepancy in this F and t relationship is due to rounding errors.

14.7 Predicting

One of the important uses of a simple regression line, as we have seen, is to obtain predictions about the dependent variable based on the value of the independent variable.

We will use the *predict* subcommand of the *regress* command to predict the value of the dependent variable in the following problem.

Ralph Farmer of the Department of Agriculture is interested in the relationship between the amount of fertilizer used and the number of bushels of wheat harvested. He collects the following information on six farmers.

x pounds of fertilizer	y bushels o wheat
100	1000
150	1250
180	1710
200	2100
222	2500

He wants to predict the number of bushels of wheat that would be harvested if 160 pounds of fertilizer were used. Put pounds of fertilizer in column c1 and bushels of wheat in column c2. The worksheet should look like the following.

	C1	C2	C3
	x	y	
1	100	1000	
2	150	1250	
3	180	1710	
4	200	2100	
5	222	2500	
6			
7			
8			

Next we will analyze the data using regression analysis and a *predict* subcommand.

```
MTB >  regress c2 1 c1;
SUBC> predict 160.
```

Regression Analysis

```
The regression equation is
y = - 402 + 12.4 x

Predictor         Coef        StDev           T         P
Constant        -401.6        348.9       -1.15     0.333
x               12.404        1.987        6.24     0.008

S = 188.5       R-Sq = 92.9%      R-Sq(adj) = 90.5%

Analysis of Variance

Source             DF           SS           MS         F         P
Regression          1      1385233      1385233     38.97     0.008
Residual Error      3       106647        35549
Total               4      1491880


Predicted Values

     Fit   StDev Fit         95.0% CI              95.0% PI
  1583.0        86.8   ( 1306.7,  1859.3)   (  922.4,  2243.6)
```

The output indicates that the predicted number of bushels is 1583 for 160 pounds of fertilizer. The output gives both the confidence and prediction intervals. The 95% confidence interval for 160 pounds of fertilizer is from 1306.7 bushels to 1859.3 bushels. The 95% prediction interval for 160 pounds of fertilizer is from 922.4 bushels to 2243.6 bushels.

14.8 Regression Examples

Example 14.8A

Healthy Hamburgers has a chain of 12 stores in northern Illinois. Sales figures and profits are shown below. Our task is to obtain a regression line for the data and predict what the profit will be if a store does $10 million in sales.

sales, x (millions)	profits, y (millions)
7.00	0.15
2.00	0.10
6.00	0.13
4.00	0.15
14.00	0.25
15.00	0.27
16.00	0.24
12.00	0.20
14.00	0.27
20.00	0.44
15.00	0.34
7.00	0.17

To do this analysis we will put sales in column c1 and name the column '*sales*'. We will put profits in column c2 and name the column '*profits*'.

We will begin our analysis by plotting the data.

```
MTB > Plot 'profit'*'sales'
```

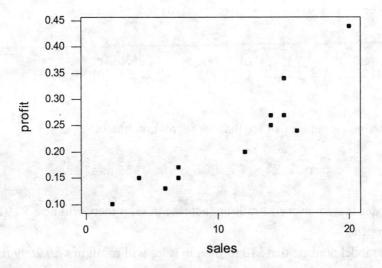

Next we will analyze the data by finding a regression line for the data.

```
MTB > regress c2 1 c1;
SUBC> predict 10.
```

Regression Analysis

```
The regression equation is
profits = 0.0506 + 0.0159 sales

Predictor        Coef       StDev          T        P
Constant      0.05060     0.02687       1.88    0.089
sales        0.015930    0.002196       7.25    0.000

S = 0.04074      R-Sq = 84.0%     R-Sq(adj) = 82.4%


Analysis of Variance

Source            DF          SS          MS         F        P
Regression         1    0.087298    0.087298     52.61    0.000
Residual Error    10    0.016594    0.001659
Total             11    0.103892


Unusual Observations
Obs      sales     profits         Fit    StDev Fit    Residual    St Resid
 10       20.0      0.4400      0.3692       0.0230      0.0708        2.11R

R denotes an observation with a large standardized residual

Predicted Values

    Fit   StDev Fit         95.0% CI              95.0% PI
 0.2099      0.0120    ( 0.1832,  0.2366)    ( 0.1153,  0.3045)
```

From the regression output we can see that the regression line is

$$\text{profits} = 0.0506 + 0.0159 \text{ sales}$$

This line indicates that for every $1 million increase in sales there will be a $0.0159 million dollar increase in profits.

The regression model predicts that $10 million in sales will result in a $0.2099 million in profits. The 95% prediction interval for this profit is from $0.1153 million to $0.3045 million.

The *regress* command calculated the *p*-value for the sales coefficient as 0. Therefore, we can conclude that the sales coefficient does not equal 0, which means that sales do explain profits.

Example 14.8B

Krugman and Rust (1987) investigated the effects of cable TV penetration on network share of advertising revenue by using simple regression analysis. Use MINITAB to find a regression line for the following data and find out the predicted network share of TV advertising for the 54 % of U.S. households that subscribe to cable.

Year	U.S. Cable TV Households (%), x	Network Share of Television Revenues, y
1980	21.1	98.9
1981	23.7	97.9
1982	25.8	96.5
1983	30.0	95.2
1984	35.7	94.0
1985	41.1	91.9

We will put the network data in column c1 and name the column '*network*'. The cable data will go in column c2, which we will name '*cable*'.

We will first plot the data.

```
MTB > plot 'network'*'cable'
```

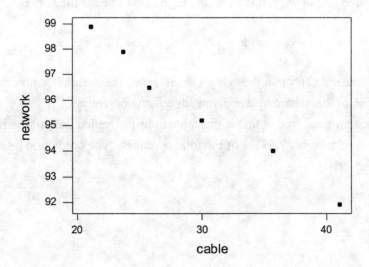

We will now regress the data set.

```
MTB > regress c1 1 c2;
SUBC> predict 54.
```

Regression Analysis

```
The regression equation is
network = 106 - 0.335 cable

Predictor        Coef        StDev          T        P
Constant      105.634        0.718     147.22    0.000
cable        -0.33486      0.02362     -14.18    0.000

S = 0.4030      R-Sq = 98.0%      R-Sq(adj) = 97.6%

Analysis of Variance

Source          DF          SS          MS         F        P
Regression       1      32.644      32.644    200.98    0.000
Residual Error   4       0.650       0.162
Total            5      33.293




Predicted Values

     Fit   StDev Fit         95.0% CI              95.0% PI
  87.551       0.600    ( 85.885,  89.218)    ( 85.544,  89.559) XX
X  denotes a row with X values away from the center
XX denotes a row with very extreme X values
```

From the regression output we can see that the regression line for the data is

$$\text{network} = 106 - 0.335 \text{ cable}$$

This regression line indicates that for every 1% increase in households with cable, the networks lose 0.335% of their share of television advertising revenues.

The regression commands predict that if the cable industry gained 54% of the households, then the networks would take in 87.551% of the total revenues. The prediction interval would be from 85.544% to 89.559%.

14.9 Statistical Summary

In this chapter we extended our use of regression analysis. We determined whether the β coefficient would be significantly different from zero or not. We tested this by seeing if the *p*-value calculated for the *b* coefficient was less than the alpha value that we were interested in. We calculated the confidence interval for β. We found the regression line for a given data set. We made predictions using the regression line and found the confidence interval for those predictions.

CHAPTER 15
MULTIPLE LINEAR REGRESSION

15.1 Introduction

In the previous two chapters we examined the relationship between two variables. In this chapter we will look at the relationship among three or more variables, using regression analysis.

When we looked at the relationship between two variables using regression analysis, one variable explained the other one, the variable of interest. Now we will have two or more variables explaining the variable of interest. The variables which are doing the explaining are called independent variables. The variable that is being explained is called the dependent variable. The mathematical formula for the regressed equation looks like the following:

$$y = a + b_1 x_1 + b_2 x_2 + b_3 x_3 + \ldots + b_n x_n$$

The x's are the independent variables. The y is the dependent variable. The b's are the coefficients of the independent variables.

A regression that contains more than one independent variable is called a multiple linear regression. We will see that many of the analyses and tests done in simple linear regression extend over to multiple linear regression.

Suppose we believe that years of education and years of work experience can explain the amount of annual salary. To test this, we have the following data.

y	$x1$	$x2$
Annual salary (thousands	Years of educatio	Years of work experienc
15	5	7
17	10	5
26	9	14
24	13	8
27	15	6

To test our belief, we will use MINITAB to do a regression analysis on the above data. First we enter the data in the worksheet, which should then look like the following.

	C1	C2	C3
	y	x1	x2
1	15	5	7
2	17	10	5
3	26	9	14
4	24	13	8
5	27	15	6
6			
7			
8			

Then we use the *regress* command and get the following results.

```
MTB > regress c1 2 c2 c3
```

Regression Analysis

```
The regression equation is
y = 0.98 + 1.24 x1 + 0.992 x2

Predictor        Coef       StDev          T        P
Constant        0.980       3.439       0.29    0.802
x1             1.2385      0.2258       5.48    0.032
x2             0.9925      0.2457       4.04    0.056

S = 1.702       R-Sq = 95.1%     R-Sq(adj) = 90.3%

Analysis of Variance

Source           DF          SS          MS        F        P
Regression        2     113.009      56.504    19.51    0.049
Residual Error    2       5.791       2.896
Total             4     118.800

Source       DF     Seq SS
x1            1     65.773
x2            1     47.236
```

This computer output shows that the regression equation is:

```
y = 0.98 + 1.24 x1 + 0.992 x2
```

This equation, which gives the relationship between annual salary and years of education and years of work experience, suggests that for every additional year of education there will be a

236   CHAPTER 15

$1,240 increase in annual salary. It also suggests that for every year of additional work experience there will be a $992 increase in annual salary.

Let us now do the regression in EXCEL. Enter the data in EXCEL as shown below.

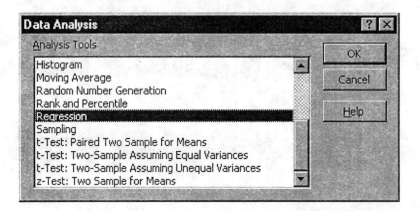

To do the regression, choose **Tools-> Data Analysis** to get the *Data Analysis* dialog box. In the *Data Analysis* dialog box, choose the *regression* option as shown below.

Then fill out the *Regression* dialog box as shown below. When finished, pressed the *OK* button.

MULTIPLE LINEAR REGRESSION   237

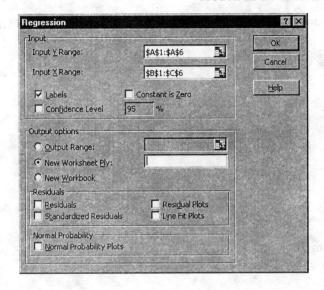

EXCEL produces the following regression report.

SUMMARY OUTPUT

Regression Statistics	
Multiple R	0.97532178
R Square	0.951252575
Adjusted R Square	0.90250515
Standard Error	1.701645392
Observations	5

ANOVA

	df	SS	MS	F	Significance F
Regression	2	113.0088059	56.50440296	19.51390409	0.048747425
Residual	2	5.791194082	2.895597041		
Total	4	118.8			

	Coefficients	Standard Error	t Stat	P-value	Lower 95%	Upper 95%	Lower 95.0%	Upper 95.0%
Intercept	0.980274745	3.438917382	0.285053299	0.802410717	-13.81620282	15.77675231	-13.81620282	15.77675231
x1	1.238464248	0.225824662	5.484185117	0.031677408	0.266818475	2.210110021	0.266818475	2.210110021
x2	0.992462135	0.245723798	4.038933725	0.056184404	-0.064802771	2.04972704	-0.064802771	2.04972704

15.2 R-Square

To see how well the independent variables (education and experience) explain the variance of the dependent variable (salary) in the preceding section, we should look at the R-square. The following part of the MINITAB output shows that the independent variables explain 95.1% of the variation of the dependent variable.

```
s = 1.702      R-sq = 95.1%      R-sq(adj) = 90.3%
```

The R-square can also be calculated using the following part of the MINITAB output.

```
Analysis of Variance

SOURCE        DF           SS           MS          F          p
Regression     2      113.009       56.504      19.51      0.049
Error          2        5.791        2.896
Total          4      118.800
```

$$R\text{-}square = \frac{113.009}{118.8} = 0.95$$

A Venn diagram of the R-square would look like the following.

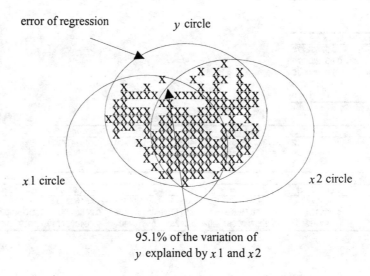

15.3 F Test

Many times in multiple regression we use sample data for regression analysis instead of population data. For this reason, we are interested in knowing whether all the true population regression slope coefficients equal zero. To test this, we will use the F test. The F statistic look like the following:

MULTIPLE LINEAR REGRESSION

$$F_{k,n-k-1} = \frac{\frac{\sum_{i=1}^{n}(\hat{y}_i - \bar{y})^2}{k}}{\frac{\sum_{i=1}^{n}(\hat{y}_i - \hat{y}_i)^2}{n-k-1}}$$

The hypotheses would be

$H_0: \beta_1 = \beta_2 = 0$
$H_1:$ at least one of β_1 or β_2 is not equal to zero

where β_1 and β_2 are population regression parameters.

Below is a reproduction of the part of the regression analysis (from Section 15.1) that we will need for the F test.

```
Analysis of Variance

SOURCE          DF          SS          MS          F          p
Regression      2           113.009     56.504      19.51      0.049
Error           2           5.791       2.896
Total           4           118.800
```

If we test the hypothesis at the 0.05 alpha level, there are two ways to see if we can reject or accept the null hypothesis.

The first way to test the null hypothesis is to find the critical value for a significant level of alpha = 0.05. In MINITAB we can use the *invcdf* command to find this value, as illustrated below:

```
MTB > invcdf .95;
SUBC> f 2 2.
```

Inverse Cumulative Distribution Function

```
F distribution with 2 DF in numerator and 2 DF in denominator

P( X <= x)            x
   0.9500          19.0000
```

If we have an alpha of 0.05, then the area of interest is $1 - 0.05 = 0.95$. The *invcdf* command works with the *f* subcommand to find the F value that corresponds with an area of 0.95. From the MINITAB output after the *invcdf* command, we can see that the area of 0.95 corresponds with an F value of 19. The first argument, '.95', of the *invcdf* indicates the area of interest. The first argument, '2', of the *f* subcommand indicates the first degree of freedom for the F test. This degree of freedom is equal to the number of independent variables. The second argument, '2', of the *f* subcommand indicates the second degree of freedom for the F test. This degree of freedom is obtained by taking the number of data items minus the number of independent variables minus one, or mathematically, $n - k - 1$. In this case we have 5 data items minus 2 independent variables minus 1 equals 2.

We will reject the null hypothesis if the calculated F value is greater than the F value that corresponds with an alpha value of 0.05.

Since the F value that corresponds with an alpha value of 0.05 is 19 and MINITAB calculated the F value as 19.51, we can conclude that at least one of the regression coefficients is significantly different from zero. This is shown below.

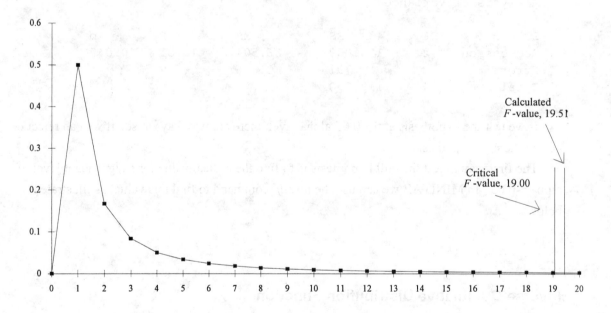

The second way to test the null hypothesis is to find the *p*-value for the calculated *F* value. In a one-tail test like the *F* test, the *p*-value is the area to the right of the calculated *F* value. MINITAB calculated the *p*-value as 0.049. This means that the area to the right of the calculated *F* value is 0.049. An alpha value of 0.05 means that there is an *F* value that has an area to the right of 0.05. We reject the null hypothesis and accept the alternative hypothesis when the *p*-value is less than the alpha value. Since our *p*-value is less than the alpha value of 0.05, we reject the null hypothesis that all the population coefficients are equal to zero.

15.4 *t* Test

From the *F* test we know that at least one of the two independent variables has a significant effect on the dependent variable. We will now use the *t* test to see if each independent variable has a significant effect.

Let us begin by testing if x_1, years of education, has a significant effect on annual salary. The *t* test statistic is the following:

$$t = \frac{\frac{\sum_{i=1}^{n}(x_i - \bar{x})(y_i - \bar{y})}{\sum_{i=1}^{n}(x_i - \bar{x})^2} - 0}{\frac{S_e}{\sqrt{\sum_{i=1}^{n}(x_i - \bar{x})^2}}} = \frac{b - 0}{S_b}$$

The null and alternative hypotheses would be

$$H_0: \beta_1 = 0$$
$$H_1: \beta_1 \neq 0$$

Below is a reproduction of the calculated *t* value from the regression output in Section 15.1. MINITAB calculates the *t* statistic as a two-tail test.

```
Predictor      Coef       Stdev      t-ratio      p
Constant       0.980      3.439      0.29         0.802
x1             1.2385     0.2258     5.48         0.032
x2             0.9925     0.2457     4.04         0.056
```

From this we can see that the calculated t statistic for x_1, or years of education, is 5.48, and the corresponding p-value 0.032. Since the p-value is less than the alpha value of 0.05, we will reject the null hypothesis that x_1 has no effect on the dependent variable. That is, we would reject the null hypothesis and accept the alternative hypothesis.

The hypotheses for analyzing the independent variable x_2, years of work experience, is:

$$H_0: \beta_2 = 0$$
$$H_1: \beta_2 \neq 0$$

From the MINITAB output we can see that the calculated p-value equals 0.056. Since the p-value is greater than the alpha value, we cannot reject the null hypothesis that this independent variable has no significant effect on the dependent variable.

Calculating t statistics is one area where we can see that using MINITAB is very easy compared to doing the statistical analysis manually, which would take a very long time.

15.5 Confidence Interval of β

We have done hypothesis testing to make inferences about the population. We can also make inferences about the population by creating confidence intervals for the regression coefficients. We can do this by using the macro we created in Chapter 14 to create confidence intervals for simple regressions, because a simple regression is actually a multiple regression with only one independent variable. Below, for convenience, is a reproduction of part of the regression output in Section 15.1.

```
Predictor         Coef        Stdev       t-ratio        p
Constant         0.980        3.439        0.29        0.802
x1               1.2385       0.2258       5.48        0.032
x2               0.9925       0.2457       4.04        0.056

s = 1.702      R-sq = 95.1%      R-sq(adj) = 90.3%
```

We will first create a 95% confidence interval for the x_1 coefficient using the *bintreg* macro.

We should remember to name the column of the independent variable that we are interested in as 'x'. We do this because we wrote the *bintreg* macro to calculate the confidence interval for the column named 'x'. The worksheet should look like the following:

MULTIPLE LINEAR REGRESSION

	C1	C2 (x)	C3
1	15	5	7
2	17	10	5
3	26	9	14
4	24	13	8
5	27	15	6
6			

Now we are ready to use the *bintreg* macro.

```
MTB > %bintreg
Executing from file: bintreg.MAC
What is the s for the confidence interval?
DATA> 1.702
What is the confidence (%) for the confidence interval?
DATA> .95
What is the beta coefficient for the confidence interval?
DATA> 1.2385
confidence interval is:
```

Data Display

```
Row    lower      upper

 1    0.534521   1.94248
```

We can see from the *bintreg* output that the confidence interval for β_1 is from 0.534521 to 1.94248.

We will now create a 95% confidence interval for the x_2 coefficient using the *bintreg* macro.

We should remember again that we would need to name the column of the independent variable that we are interested in as 'x'. The worksheet should look like the following:

Row	C1	C2	C3
			x
1	15	5	7
2	17	10	5
3	26	9	14
4	24	13	8
5	27	15	6
6			

Now we can use the *bintreg* macro to find the confidence interval for the second independent variable.

```
MTB > %bintreg
Executing from file: bintreg.MAC
What is the s for the confidence interval?
DATA> 1.702
What is the confidence (%) for the confidence interval?
DATA> .95
What is the beta coefficient for the confidence interval?
DATA> .9925
confidence interval is:
```

Data Display

```
Row     lower     upper

 1    0.226488   1.75851
```

We can see from the *bintreg* output that the confidence interval for β_2 is from 0.226488 to 1.75851.

15.6 Predicting

If we look more carefully at the regression equation for the annual salary problem, y = 0.98 + 1.24 x1 + 0.992 x2 , we realize that this equation is also a model of the relationship between all annual salaries and the years of experience and years of education. We can use this model to *predict* the value of annual salaries, *y*, which we have not observed.

Suppose we are interested in predicting what the annual salary would be for someone who has 12 years of education and 12 years of work experience. As in Chapter 14, we will use the *predict* subcommand of the *regress* command.

```
MTB > regress c1 2 c2 c3;
SUBC> predict 12 12.
```

Regression Analysis

```
The regression equation is
y = 0.98 + 1.24 x1 + 0.992 x2

Predictor         Coef        StDev          T        P
Constant         0.980        3.439       0.29    0.802
x1              1.2385       0.2258       5.48    0.032
x2              0.9925       0.2457       4.04    0.056

S = 1.702       R-Sq = 95.1%      R-Sq(adj) = 90.3%

Analysis of Variance

Source             DF           SS          MS          F        P
Regression          2      113.009      56.504      19.51    0.049
Residual Error      2        5.791       2.896
Total               4      118.800

Source          DF     Seq SS
x1               1     65.773
x2               1     47.236


Predicted Values

     Fit   StDev Fit        95.0% CI              95.0% PI
  27.751       1.349   ( 21.948,  33.555)   ( 18.409,  37.094)
```

From the above MINITAB output we can see that the predicted annual salary is $27,751 and the 95% prediction interval is from $18,409 to $37,094.

15.7 Another Regression Example

To show how the multiple regression technique can be used by real estate appraisers, Andres and Ferguson (1986) use the data below to do a multiple regression analysis. They used this formula:

$$y_i = a + b_1 x_{1i} + b_2 x_{2i} + e_i$$

where

y_i = sale price for the ith house
x_{1i} = home size for the ith house
x_{2i} = condition rating for the ith house

Sale Price, y (thousands of dollars)	Home Size, x1 (hundreds of sq. ft.)	Condition Rating, x2 (1 to 10)
60.0	23	5
32.7	11	2
57.7	20	9
45.5	17	3
47.0	15	8
55.3	21	4
64.5	24	7
42.6	13	6
54.5	19	7
57.5	25	2

Use MINITAB to do a regression on the data and predict what the sale price of a house would be if the home size was 1800 square feet and the condition rating was 5.

To do this we will put the sale price data in column c1 and name column c1 '*price*'. We will put the home size data in column c2 and name column c2 '*size*'. We will put the rating data in column c3 and name column c3 '*rating*'.

```
MTB > regress c1 2 c2 c3;
SUBC> predict 18 5.
```

Regression Analysis

The regression equation is
price = 9.78 + 1.87 size + 1.28 rating

```
Predictor       Coef        StDev         T        P
Constant       9.782        1.630      6.00    0.001
size         1.87094      0.07617     24.56    0.000
rating        1.2781       0.1444      8.85    0.000

S = 1.081      R-Sq = 99.0%     R-Sq(adj) = 98.7%
```

Analysis of Variance

```
Source           DF         SS         MS         F        P
Regression        2      819.33     409.66    350.87    0.000
Residual Error    7        8.17       1.17
Total             9      827.50

Source       DF     Seq SS
size          1     727.85
rating        1      91.48
```

Unusual Observations
```
Obs     size     price       Fit    StDev Fit   Residual    St Resid
 10     25.0    57.500    59.112       0.766     -1.612      -2.11R
```

R denotes an observation with a large standardized residual

Predicted Values
```
    Fit   StDev Fit       95.0% CI            95.0% PI
 49.850       0.349   ( 49.023,  50.677)  ( 47.163,  52.537)
```

From the MINITAB output we can see that the regression equation is

$$\text{price} = 9.78 + 1.87 \text{ size} + 1.28 \text{ rating}$$

This regression equation says that for every 100 square foot increase in size, a house's price will increase by \$1,870. For every increase in rating, there will be a \$1,290 increase in the price of a house.

From the regression output we can see that the coefficients for size and rating do not equal zero at the 5% alpha level. We conclude this because the p-value calculated for both coefficients is zero and this would be less than the 5% alpha level.

The regression predicts that the sale price would be \$49,850 for a house with 1800 square feet and a rating of 5. The confidence interval for this prediction is from \$47,164 to \$52,536.

15.8 Stepwise Regression

In regression analysis, a large number of variables can contribute to the explanation of the dependent variable. Some of the independent variables will make a large contribution and some only a small contribution to the explanation of the dependent variable. In regression analysis we are only interested in the independent variables that make a large or significant contribution to the explanation of the dependent variable.

Stepwise regression is a method for determining which of the possible independent variables will make a significant contribution to the explanation of the dependent variable. Stepwise regression involves the following steps.

1. First, we run a simple regression on each explanatory variable and then choose the model that explains the highest amount of variation in the dependent variable. This will be the regression with the highest R-square value.
2. The second variable we enter should be the one that, in conjunction with the first variable, explains the greatest amount of variation of the dependent variable. Before entering this variable we run an F test to see if the increase in R-square is significant.
3. We keep on adding independent variables until there is no significant increase in R-square.

We will illustrate this process by doing a stepwise regression on the problem in Section 15.7. The format of the worksheet should be the same.

```
MTB > stepwise c1 c2 c3
```

Stepwise Regression

```
F-to-Enter:        4.00    F-to-Remove:        4.00

Response is  price   on  2 predictors, with N =   10

    Step           1         2
Constant      16.008     9.782

size           1.900     1.871
T-Value         7.64     24.56

rating                    1.28
T-Value                   8.85

S              3.53      1.08
R-Sq          87.96     99.01
 More? (Yes, No, Subcommand, or Help)
SUBC> no
```

The first argument of the *stepwise* command, 'c1', is the location of the dependent variable of interest. The second argument, 'c2', is the location of the first independent variable to be considered in our regression model. The third argument, 'c3', is the location of the second independent variable to be considered.

The second column of the worksheet, *'size'*, is thus the first significant variable chosen. The third column, *'rating'*, is the second variable chosen. From the MINITAB output, we can therefore conclude that both *'size'* and *'rating'* variables make significant contributions to explaining the sale price of a house.

15.9 Statistical Summary

In this chapter we looked at the relationship among three or more variables. We saw that many of the concepts of the relationship between two variables extend to the situations of three or more variables. We also looked at stepwise regression, a way to determine which of the independent variables made a significant contribution to the dependent variable.

CHAPTER 16
OTHER TOPICS IN APPLIED REGRESSION ANALYSIS

16.1 Introduction

When we do regression analysis there are six assumptions.

1. The dependent and independent variables have a linear relationship.
2. The expected value of the residual term is zero.
3. The variance of the residual term is constant.
4. The residual terms are independent.
5. The independent variables are uncorrelated.
6. The residuals are normally distributed.

To make the proper statistical inferences, these six assumptions must not be materially violated. In this chapter we will look at the six major assumptions by studying an article that was published in *Accounting Review*.[1] G. J. Benston did a multiple regression that had the following dependent variable and independent variables:

y = hours of labor in ith week
$x1$ = thousands of pounds shipped
$x2$ = percentage of units shipped by truck
$x3$ = average number of pounds per shipment

From the above, we can conclude that Benston felt that thousands of pounds shipped, percentage of units shipped by truck, and average number of pounds per shipment could explain the hours of labor in a particular week. The data for his model are displayed below.

---

[1] G. J. Benston (1966), "Multiple Regression Analysis of Cost Behavior," *Accounting Review* Vol. 41, No. 4, pp. 657-672.

week	y	x1	x2	x3
1	100	5.1	90	20
2	85	3.8	99	22
3	108	5.3	58	19
4	116	7.5	16	15
5	92	4.5	54	20
6	63	3.3	42	26
7	79	5.3	12	25
8	101	5.9	32	21
9	88	4.0	56	24
10	71	4.2	64	29
11	122	6.8	78	10
12	85	3.9	90	30
13	50	3.8	74	28
14	114	7.5	89	14
15	104	4.5	90	21
16	111	6.0	40	20
17	110	8.1	55	16
18	100	2.9	64	19
19	82	4.0	35	23
20	85	4.8	58	25

y = hours of labor in ith week
x1 = thousnads of pounds shipped
x2 = percentage of units shipped by truck
x3 = average number of pounds per shipmen

Let us do a regression analysis on the above data. The MINITAB worksheet should look like the following after entering the data.

	C1 week	C2 labor	C3 thousand	C4 percent	C5 average	C6	C7	C8	C9
1	1	100	5.1	90	20				
2	2	85	3.8	99	22				
3	3	108	5.3	58	19				
4	4	116	7.5	16	15				
5	5	92	4.5	54	20				
6	6	63	3.3	42	26				
7	7	79	5.3	12	25				
8	8	101	5.9	32	21				
9	9	88	4.0	56	24				
10	10	71	4.2	64	29				
11	11	122	6.8	78	10				
12	12	85	3.9	90	30				
13	13	50	3.8	74	28				
14	14	114	7.5	89	14				
15	15	104	4.5	90	21				
16	16	111	6.0	40	20				
17	17	110	8.1	55	16				
18	18	100	2.9	64	19				
19	19	82	4.0	35	23				
20	20	85	4.8	58	25				

We will now regress the above data using the *regress* command.

252   CHAPTER 16

```
MTB > brief 3
MTB > regress c2 3 c3-c5, store st. residual c7, fits c8;
SUBC> residual c9;
SUBC> dw.
```

Regression Analysis

The regression equation is
labor = 132 + 2.73 thousand + 0.0472 percent - 2.59 average

```
Predictor        Coef        StDev          T         P
Constant       131.92        25.69       5.13     0.000
thousand        2.726        2.275       1.20     0.248
percent       0.04722      0.09335       0.51     0.620
average       -2.5874       0.6428      -4.03     0.001

S = 9.810       R-Sq = 77.0%      R-Sq(adj) = 72.7%
```

Analysis of Variance

```
Source             DF          SS          MS         F         P
Regression          3      5158.3      1719.4     17.87     0.000
Residual Error     16      1539.9        96.2
Total              19      6698.2

Source             DF      Seq SS
thousand            1      3400.6
percent             1       198.4
average             1      1559.3
```

```
Obs   thousand     labor        Fit   StDev Fit    Residual    St Resid
  1       5.10    100.00      98.33       3.53        1.67        0.18
  2       3.80     85.00      90.03       4.40       -5.03       -0.57
  3       5.30    108.00      99.95       2.50        8.05        0.85
  4       7.50    116.00     114.31       5.47        1.69        0.21
  5       4.50     92.00      94.99       3.12       -2.99       -0.32
  6       3.30     63.00      75.63       4.08      -12.63       -1.42
  7       5.30     79.00      82.25       5.13       -3.25       -0.39
  8       5.90    101.00      95.18       3.46        5.82        0.63
  9       4.00     88.00      83.37       2.80        4.63        0.49
 10       4.20     71.00      71.36       4.39       -0.36       -0.04
 11       6.80    122.00     128.27       5.71       -6.27       -0.79
 12       3.90     85.00      69.18       5.63       15.82        1.97
 13       3.80     50.00      73.33       3.92      -23.33       -2.59R
 14       7.50    114.00     120.35       5.48       -6.35       -0.78
 15       4.50    104.00      94.10       3.53        9.90        1.08
 16       6.00    111.00      98.42       3.04       12.58        1.35
 17       8.10    110.00     115.20       5.35       -5.20       -0.63
 18       2.90    100.00      93.69       6.40        6.31        0.85
 19       4.00     82.00      84.97       3.94       -2.97       -0.33
 20       4.80     85.00      83.06       2.92        1.94        0.21
```

R denotes an observation with a large standardized residual

Durbin-Watson statistic = 2.43

We will name column c8 *'fits'* and column c9 *'residual'*.

The worksheet should look like the following after the *regress* command.

	C1	C2	C3	C4	C5	C6	C7	C8	C9
	week	labor	thousand	percent	average			fits	residual
1	1	100	5.1	90	20		0.18267	98.328	1.6719
2	2	85	3.8	99	22		-0.57430	90.034	-5.0343
3	3	108	5.3	58	19		0.84869	99.950	8.0502
4	4	116	7.5	16	15		0.20709	114.314	1.6862
5	5	92	4.5	54	20		-0.32171	94.993	-2.9926
6	6	63	3.3	42	26		-1.41567	75.630	-12.6300
7	7	79	5.3	12	25		-0.38907	82.253	-3.2531
8	8	101	5.9	32	21		0.63374	95.183	5.8172
9	9	88	4.0	56	24		0.49204	83.374	4.6258
10	10	71	4.2	64	29		-0.04102	71.360	-0.3599
11	11	122	6.8	78	10		-0.78618	128.270	-6.2703
12	12	85	3.9	90	30		1.96944	69.182	15.8177
13	13	50	3.8	74	28		-2.59435	73.329	-23.3291
14	14	114	7.5	89	14		-0.77991	120.348	-6.3481
15	15	104	4.5	90	21		1.08116	94.105	9.8950
16	16	111	6.0	40	20		1.34863	98.421	12.5794
17	17	110	8.1	55	16		-0.63291	115.203	-5.2035
18	18	100	2.9	64	19		0.84816	93.690	6.3095
19	19	82	4.0	35	23		-0.33057	84.970	-2.9700
20	20	85	4.8	58	25		0.20694	83.062	1.9379

From the MINITAB output we can see that the regressed equation is

```
labor = 132 + 2.73 thousand + 0.0472 percent - 2.59 average
```

This equation tells us that

1. For every thousand pounds' increase, there will be 2.73 hours' increase in labor.
2. For every 1% increase in units shipped, there will be 0.0472 hour's increase in labor.
3. For every average pound increase per shipment, there will be 2.59 hours' decrease in labor.

From the R-square calculation of 72.7% we can see that the independent variables explain 72.7% of the variability of the dependent variable.

Let us now do the regression in EXCEL. Input the data in EXCEL as shown below.

254   CHAPTER 16

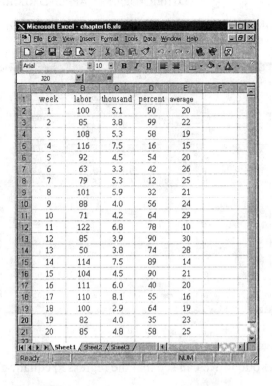

Choose **Tools->Data Analysis** to get the *Data* Analysis dialog box. Choose the *regression* option as shown below.

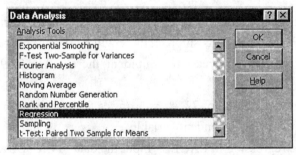

Fill out the *Regression* dialog box as shown below.

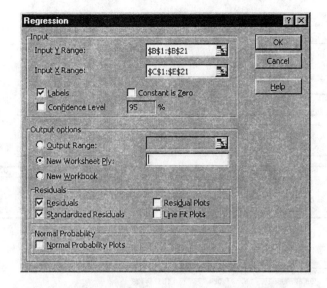

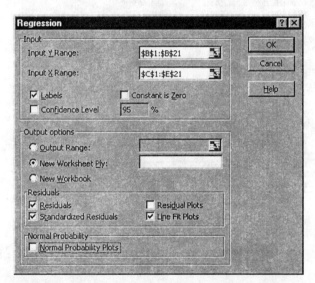

After pressing *OK* on the *Regression* dialog box, the following EXCEL regression report is produced.

SUMMARY OUTPUT

Regression Statistics	
Multiple R	0.87755597
R Square	0.77010448
Adjusted R Square	0.72699907
Standard Error	9.810345853
Observations	20

ANOVA

	df	SS	MS	F	Significance F
Regression	3	5158.313828	1719.437943	17.86561083	2.32332E-05
Residual	16	1539.886172	96.24288576		
Total	19	6698.2			

	Coefficients	Standard Error	t Stat	P-value	Lower 95%	Upper 95%	Lower 95.0%	Upper 95.0%
Intercept	131.9242521	25.69321439	5.134595076	9.98597E-05	77.45708304	186.3914211	77.45708304	186.3914211
thousand	2.72608977	2.275004884	1.198278645	0.24825743	-2.096704051	7.548883591	-2.096704051	7.548883591
percent	0.047218412	0.093348559	0.505829045	0.6198742	-0.150671647	0.245108472	-0.150671647	0.245108472
average	-2.587443905	0.642818185	-4.025156669	0.000978875	-3.950157275	-1.224730536	-3.950157275	-1.224730536

RESIDUAL OUTPUT

Observation	Predicted labor	Residuals	Standard Residuals
1	98.32808892	1.671911084	0.185714438
2	90.03425012	-5.034250116	-0.559200151
3	99.94976158	8.05023842	0.894213524
4	114.3137614	1.686238625	0.187305929
5	94.99257221	-2.992572209	-0.332412334
6	75.62998011	-12.62998011	-1.402927271
7	82.25305118	-3.25305118	-0.361346113
8	95.18284891	5.81715109	0.646164114
9	83.37418853	4.625811472	0.513831139
10	71.35993426	-0.359934255	-0.039981186
11	128.2702596	-6.270259628	-0.696495019
12	69.18234214	15.81765786	1.75701176
13	73.32912638	-23.32912638	-2.591379189
14	120.3481494	-6.348149382	-0.705146945
15	94.10499115	9.895008851	1.099129028
16	98.42064909	12.57935091	1.397303424
17	115.2034894	-5.203489414	-0.577999105
18	93.69045661	6.309543394	0.700858625
19	84.97004577	-2.970045774	-0.329910117
20	83.06205326	1.937946736	0.215265448

16.2 Linearity

One way to check linearity is to plot the y variable with each x variable to check if the plot looks linear. If one of these plots does not look linear, then the linearity assumption might be violated. Below are three plots of y with each of the three x variables. For each plot we will also look at the strength of the relationship between y and x.

```
MTB > gstd
MTB > plot 'labor' 'thousand'
```
Plot

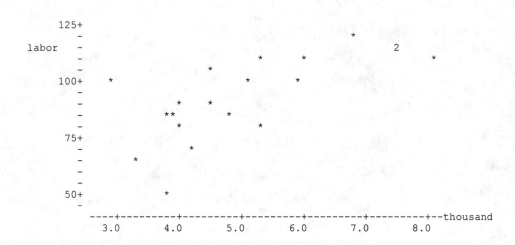

```
MTB > correlation 'labor' 'thousand'
```

Correlations (Pearson)

Correlation of labor and thousand = 0.713, P-Value = 0.000

```
MTB > plot 'labor' 'percent'
```
Plot

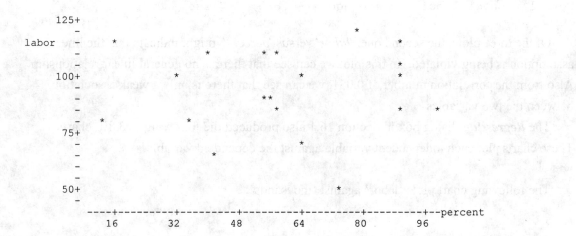

```
MTB > correlation 'labor' 'percent'
```
Correlations (Pearson)

```
Correlation of labor and percent = 0.031, P-Value = 0.898

MTB > plot 'labor' 'average'
```
Plot

```
       125+
          -        *
 labor    -            *   *
          -              *        * *
          -                *         *
       100+                    * * *
          -                  *
          -                      *    *
          -                         *  *        *
       75+
          -                                        *
          -
          -                              *
       50+
          -                                  *
           --------+---------+---------+---------+---------+--------average
                 12.0      16.0      20.0      24.0      28.0
```

```
MTB > correlation 'labor' 'average'
```

Correlations (Pearson)

Correlation of labor and average = -0.865, P-Value = 0.000

Of the three plots, the second one, *'labor'* versus *'percent'*, might indicate that the linear assumption is being violated. In this plot we can see that there is no general linear relationship. Also from the correlation number of 0.031, we can see that there is only a weak association between the two variables.

The *Regression* dialog box in Section 16.1 also produced the following EXCEL charts. These charts plot each independent variable against the dependent variable.

The following chart plots 'labor' against 'thousands'.

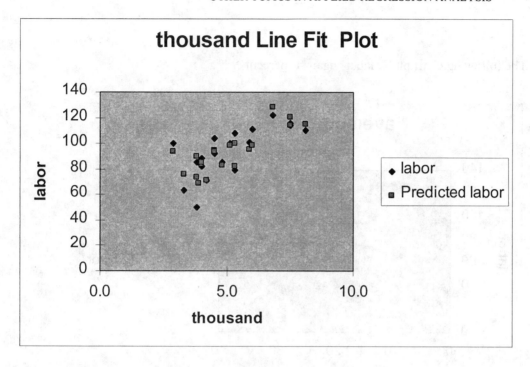

The following chart plots 'labor' against 'percent'.

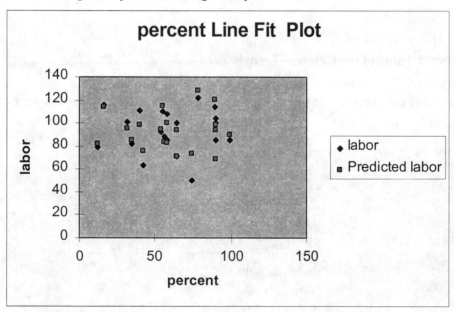

The following chart plots 'labor' against 'percent'.

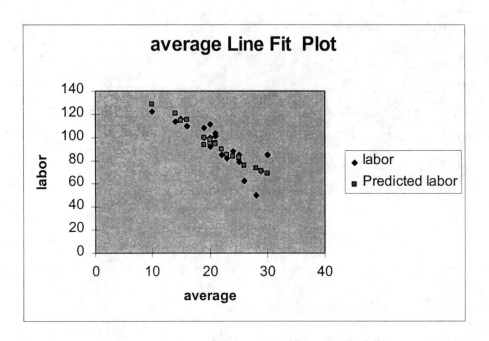

16.3 The Expected Value of the Residual Term is Zero

One way to check this assumption is to produce a dotplot and a boxplot of the residual terms and see if the midpoint is zero.

```
MTB > dotplot 'residual'
```

Dotplot

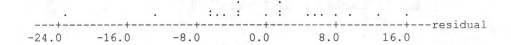

```
MTB > boxplot 'residual'
```

Boxplot

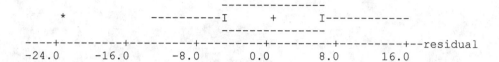

From the above dotplot and boxplot we can see that the midpoint is very close to zero.

16.4 The Variance of the Error Term Is Constant

One quick way to check to see if the variance of the error term is constant is to see if the plot of the residuals looks like the following.

Heteroscedasticity

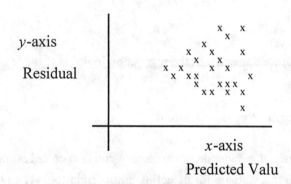

If the plot looks like the above, then the assumption that the variance of the error terms is constant is violated. This problem is called *heteroscedasticity*.

```
MTB > plot 'residual' 'fits'

Plot
```

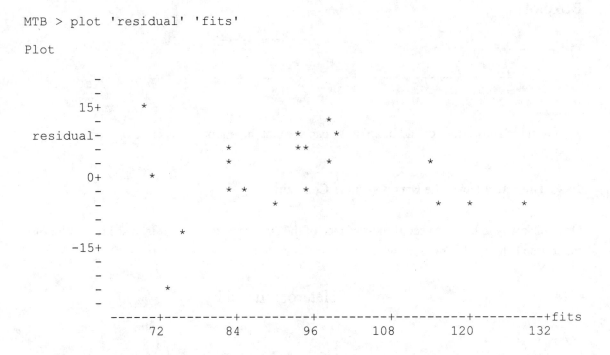

From the above plot we can see that the regression looks as if it has constant variance.

16.5 The Residual Terms Are Independent

If the residual terms are not independent, we have a situation called *autocorrelation*. The Durbin-Watson test is a test statistic for detecting autocorrelation. The Durbin-Watson calculation from the MINITAB regression output is reproduced below.

```
Durbin-Watson statistic = 2.43
```

The DW statistic has a value between 0 and 4. A 0 means very strong positive autocorrelation. A 4 means very strong negative autocorrelation. A value of 2 means no or very little autocorrelation. Our test statistic of 2.43, therefore, suggests that our data set has very little autocorrelation.

According to James Stevens:

> This is a very important assumption. . . . . if independence is violated only mildly then the probability of a type 1 error will be *several* times greater than the level the experimenter thinks he or she is working at. Thus, instead of rejecting falsely 5% of the time, the experimenter may be rejecting falsely 25% or 30% of the time.[2]

The regression analysis in EXCEL did not calculate the Durbin-Watson statistics. We can calculate it easily by using EXCEL regression report done in Section 16.1. We will create the DW statistics based on its definition. The definition of the DW statistics is

$$DW = \frac{\sum_{t=2}^{n}(e_t - e_{t-1})^2}{\sum_{t=1}^{n}e_t^2}$$

We will use the residual column of regression report to calculate our DW statistics. The *SUMXMY2* EXCEL function to calculate the numerator of the DW statistics, the SUMSQ EXCEL function to calculate denominator of the DW statistics. This is shown below.

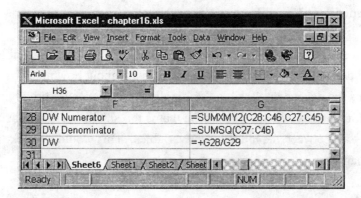

The DW calculation is shown below.

---

[2] James Stevens, *Applied Multivariate Statistics for the Social Sciences* (Hillsdale, N.J.: Lawrence Erlbaum Associates, 1986).

[Excel spreadsheet screenshot showing residual analysis with the following data:]

Observation	Predicted labor	Residuals	Standard Residuals		
1	98.32808892	1.671911084	0.185714438		
2	90.03425012	-5.034250116	-0.559200151	DW Numerator	3741.576872
3	99.94976158	8.05023842	0.894213524	DW Denominator	1539.886172
4	114.3137614	1.686238625	0.187305929	DW	2.429774966
5	94.99257221	-2.992572209	-0.332412334		
6	75.62998011	-12.62998011	-1.402927271		
7	82.25305118	-3.25305118	-0.361346113		
8	95.18284891	5.81715109	0.646164114		
9	83.37418853	4.625811472	0.513831139		
10	71.35993426	-0.359934255	-0.039981186		
11	128.2702596	-6.270259628	-0.696495019		
12	69.18234214	15.81765786	1.75701176		
13	73.32912638	-23.32912638	-2.591379189		
14	120.3481494	-6.348149382	-0.705146945		
15	94.10499115	9.895008851	1.099129028		
16	98.42064909	12.57935091	1.397303424		
17	115.2034894	-5.203489414	-0.577999105		
18	93.69045661	6.309543394	0.700858625		
19	84.97004577	-2.970045774	-0.329910117		
20	83.06205326	1.937946736	0.215265448		

16.6 The Independent Variables Are Uncorrelated

The next thing we would like to see is how correlated the independent variables are. The more correlated the independent variables, the less sure we are about how much each independent variable explains the dependent variable. Multicollinearity can cause problems like, inflating regression coefficients, producing a t statistic that is too small.[3] The ideal situation would be no correlation at all between the independent variables. This would look like the following, using a Venn diagram.

---

[3] Alan Kvanli, C. Guynes, and Robert Pavur, *Introduction to Business Statistics: A Computer Integrated Approach* (New York: West Publishing Company, 1992).

OTHER TOPICS IN APPLIED REGRESSION ANALYSIS   265

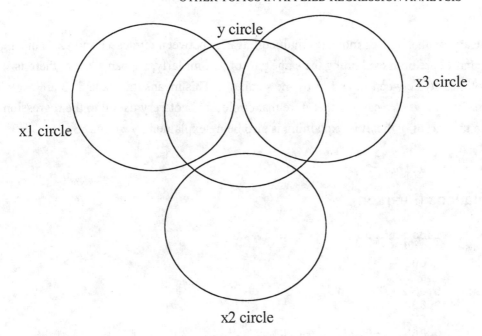

In this diagram none of the independent variables intersect any other independent variable. But most of the time this is not the case. Usually there are intersections among the independent variables, indicating some degree of correlation. This would look like the following in a Venn diagram.

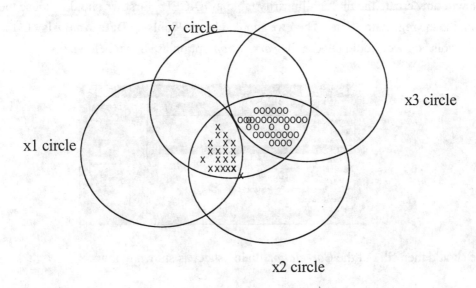

The area with small x's indicates multicollinearity between circles $x1$ and $x2$. This means that $x1$ and $x2$ are both explaining the same part of y. Similarly, we can see that there is multicollinearity between $x2$ and $x3$ by the small o's. This means that $x2$ and $x3$ are explaining the same part of y. We can see therefore that circle $x2$ is not very useful in the regression because the part of y that it is explaining is also being explained by $x1$ and $x3$.

```
MTB > correlation c2-c5
```

Correlations (Pearson)

```
          labor  thousand   percent
thousand  0.713
          0.000

percent   0.031    -0.194
          0.898     0.413

average  -0.865    -0.717    -0.013
          0.000     0.000     0.957

Cell Contents: Correlation
               P-Value
```

From this MINITAB multicollinearity table we can see that there is strong multicollinearity between thousands of pounds shipped and the average number of pounds per shipment.

We will now create the multicollinearity table in EXCEL. First we should choose the sheet that contains the regression data. Then we should choose **Tools -> Data Analysis**. In the Data Analysis tools box we should choose the *correlation* option as shown below.

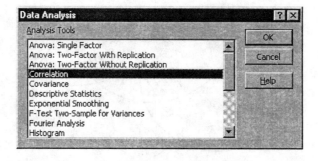

We should then fill out the *Correlation* dialog sheet as shown below.

OTHER TOPICS IN APPLIED REGRESSION ANALYSIS   267

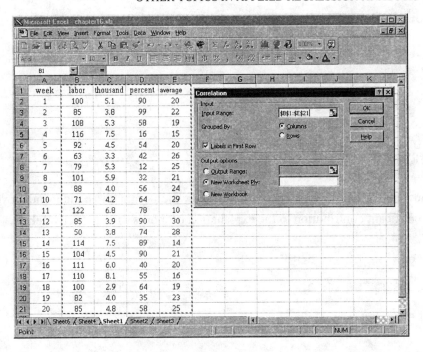

The correlation dialog box produces the following report,

	labor	thousand	percent	average
labor	1			
thousand	0.712527	1		
percent	0.030716	-0.193825	1	
average	-0.865498	-0.717235	-0.012804	1

16.7 The Residuals Are Normally Distributed

In Section 16.3, we checked to see if the expected value of the residual term is zero or not. We created a dotplot, which is reproduced below. We can see that the dotplot looks like a normal distribution.

```
MTB > dotplot 'residual'
```

Dotplot

```
                                     .    .
              .              :.. :  . :  ... .  .   .
           ---+---------+---------+---------+---------+---------+---residual
           -24.0     -16.0
```

16.8 Stepwise Regression

We will now perform a stepwise regression to see which of the three independent variables makes a significant contribution to the explanation of hours of labor per week.

```
MTB > stepwise c2 c3 c4 c5

 STEPWISE REGRESSION OF   labor   ON   3 PREDICTORS, WITH N =    20

      STEP         1
CONSTANT      160.6

average       -3.15
T-RATIO       -7.33

S              9.66
R-SQ          74.91
 MORE? (YES, NO, SUBCOMMAND, OR HELP)
SUBC> no
```

From the stepwise regression, we can see that the only variable that makes a significant contribution to the explanation of hours of labor in a week is the average number of pounds per shipment.

16.9 Statistical Summary

In this chapter we analyzed the six basic assumptions of linear regression. Analyzing them is important in every regression situation because when we make inferences about the regressed equation we are assuming that the six basic assumptions are not materially violated. If they are violated and we are not aware of that, then our inferences could be seriously flawed.

The six assumptions were
1. The dependent and independent variables have a linear relationship.
2. The expected value of the residual term is zero.
3. The variance of the residual term is constant.
4. The residual terms are independent.
5. The independent variables are uncorrelated.
6. The residuals are normally distributed.

We checked assumption 1 by plotting each individual independent variable with the dependent variable to see if the plot looked linear.

We checked assumption 2 by seeing if the dotplot and boxplot of the residuals have midpoints around 0.

We checked assumption 3 by plotting the variance of the residuals to see if the variance was constant or not.

We checked assumption 4 by calculating the Durbin-Watson statistic to see if there was any autocorrelation.

We checked assumption 5 by calculating the correlation of the independent variables.

We checked assumption 6 by dotplotting the residuals to see if the dotplot looked like a normal distribution or not.

CHAPTER 17
NONPARAMETRIC STATISTICS

17.1 Introduction

In most of the previous chapters we have been interested in estimating distribution parameters like the mean and variance and have assumed that we knew which distribution we were dealing with. In nonparametric statistics, these objectives and assumptions have been changed or cannot be achieved. Now we will be doing the test without making restrictive distribution assumptions about the parameters of the populations.

Below is a model showing how we will proceed in studying nonparametric statistics.

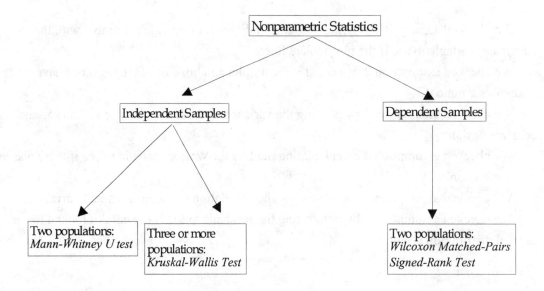

In general, relative rankings are used to test whether the probability distributions are equal or not. We will first look at a test called the Wilcoxon rank sum test, also known as the Mann-Whitney U test.

17.2 Mann-Whitney U Test

A Mann-Whitney U test is a nonparametric test of independent samples from two populations when we cannot assume that the sample is from a population that has a normal distribution.

Example 17.2A

A company wishes to compare typing accuracy on two kinds of computer keyboards. Fifteen experienced typists type the same 600 words. Keyboard A is used 7 times and keyboard B 8 times, with the following results:

	Number of Errors						
Board A	13	9	16	15	10	11	12
Board B	15	9	18	12	14	17	20	19

Use the Mann-Whitney U test to see if the two kinds of computer keyboards have the same type of accuracy at a 0.05 alpha level.

In previous chapters we used the t test for problems like this, but we had to assume that we were sampling from a normal distribution. If we assume that this keyboard problem does not follow a normal distribution, we cannot use a t test. We will use the Mann-Whitney U nonparametric test instead because it does not assume that the sample will come from a normal distribution.

Our null and alternative hypotheses are

H_0: Accuracy on the two types of keyboards is equal
H_1: Accuracy on the two types of keyboards is not equal

We will use the *mann-whitney* MINITAB command to solve this problem. We will put the board A data in column c1 and board B data in column c2 of the MINITAB worksheet.

```
MTB > mann-whitney c1 c2

Mann-Whitney Confidence Interval and Test

C1           N =    7     Median =         12.000
C2           N =    8     Median =         16.000
Point estimate for ETA1-ETA2 is      -3.500
95.7 Percent CI for ETA1-ETA2 is (-6.998,1.000)
W = 41.5
Test of ETA1 = ETA2  vs  ETA1 not = ETA2 is significant at 0.1052
The test is significant at 0.1043 (adjusted for ties)

Cannot reject at alpha = 0.05
```

From the MINITAB output we can see that we cannot reject the hypothesis that the two keyboards have equal accuracy at an alpha value of 0.05 because the p-value is 0.1052.

Example 17.2B

The table below shows the research and development expenditures of 15 companies in each of two major industries, A and B. Use the Mann-Whitney U test to see if the two industries' expenditures are equal.

R&D Expenditures of Two Major Industries (in millions of dollars)	
Industry A	Industry B
40	8
41	6
43	15
46	16
47	17
53	18
55	19
56	20
61	21
63	22
64	25
68	26
79	27
80	29
85	30

Our null and alternative hypotheses would be

H_0: The R&D expenditures of the two industries are equal
H_1: The R&D expenditures of the two industries are not equal

We put the data on industry A in column c1 and the data on industry B in column c2 of the MINITAB worksheet. The *mann-whitney* command is illustrated below.

```
MTB > mann-whitney c1 c2
```

Mann-Whitney Confidence Interval and Test

```
C1           N =  15    Median =        56.00
C2           N =  15    Median =        20.00
Point estimate for ETA1-ETA2 is         37.00
95.4 Percent CI for ETA1-ETA2 is (28.00,48.00)
W = 345.0
Test of ETA1 = ETA2  vs  ETA1 not = ETA2 is significant at 0.0000
```

From this MINITAB output, we can see that the calculated *p*-value is 0.0000. Since this is less than the alpha value of 0.05, we will reject the null hypothesis that the two industries spend the same amount on R&D.

Example 17.2C

The producer commodity price indexes for January 1985 and January 1986 for 6 product categories are shown below. The data are from *Standard & Poor's Statistical Service, Currency Statistics*, Jan. 1987, pp. 12-13. Use the Mann-Whitney U test to see if the probability distribution of these economic indexes was the same in January 1985 and January 1986.

Product Category	Jan-85	Jan-86
Processed poultry	198.80	192.40
Concrete ingredients	331.00	339.00
Lumber	343.00	329.60
Gas fuels	1,073.00	1,034.30
Drugs and pharmaceuticals	247.40	265.90
Synthetic fibers	157.60	151.10

Our null and alternative hypotheses are

H_0: The economic indexes are the same in January 1985 and January 1986.
H_1: The economic indexes are not the same in January 1985 and January 1986.

We will put the January 1985 data in column c1 and the January 1986 data in column c2.

```
MTB > mann-whitney c1 c2
```

Mann-Whitney Confidence Interval and Test

```
C1          N =   6      Median =       289.2
C2          N =   6      Median =       297.8
Point estimate for ETA1-ETA2 is         5.2
95.5 Percent CI for ETA1-ETA2 is (-181.4,192.0)
W = 41.0
Test of ETA1 = ETA2  vs  ETA1 not = ETA2 is significant at 0.8102

Cannot reject at alpha = 0.05
```

MINITAB calculates the *p*-value as 0.8102. Since this is greater than an alpha value of 0.05, we cannot reject the null hypothesis that the economic indexes are the same in January 1985 and January 1986.

17.3 Kruskal-Wallis Test

The Kruskal-Wallis test is a one-factor analysis of variance by ranks. It is a nonparametric test that represents a generalization of the two-sample Mann-Whitney *U* test to situations where more than two populations are involved. Unlike one-factor analyses of variance, the Kruskal-Wallis test makes no assumptions about the population distributions.

Example 17.3A

Shaving Cream Sales (cases per thousand of population)		
A	B	C
38	26	40
42	30	36
27	18	32
60	42	37
36	24	42
54	30	46
40	26	38

The manufacturer of a new shaving cream tests 3 new advertising campaigns in a total of 21 markets. Sales in the third week after introduction are given in the above table.

Our task is to determine whether the median sales levels for the 3 campaigns are different at the 10% alpha level. For this we will use the *kruskal-wallis* command.

To use the *kruskal-wallis* MINITAB command, we have to input the data in a special way. The data will have to go into column c1 and the group for each data in column c2. To do this we will represent group A with a 1, group B with a 2, and group C with a 3. After entering the data the worksheet should look like the following.

	C1	C2
1	38	1
2	42	1
3	27	1
4	60	1
5	36	1
6	54	1
7	40	1
8	26	2
9	30	2
10	18	2
11	42	2
12	24	2
13	30	2
14	26	2
15	40	3
16	36	3
17	32	3
18	37	3
19	42	3
20	46	3
21	38	3

Our hypotheses are

H_0: The different advertising campaigns produce no difference in sales.
H_1: The different advertising campaigns do produce different sales.

```
MTB > kruskal-wallis c1 c2

Kruskal-Wallis Test

Kruskal-Wallis Test on C1

C2            N    Median    Ave Rank         Z
1             7     40.00        14.2      1.68
2             7     26.00         5.7     -2.76
3             7     38.00        13.1      1.08
Overall      21                  11.0

H = 7.74   DF = 2   P = 0.021
H = 7.78   DF = 2   P = 0.020 (adjusted for ties)
```

From the MINITAB output we can see that the calculated *p*-value is 0.021. Since this is less than the alpha value of 0.10, we can reject the null hypothesis that there is no difference among the different advertising campaigns and accept the alternative hypothesis that the different advertising campaigns do produce different sales levels.

Example 17.3B

Assume that samples of executive vice presidents in a certain industry were drawn from firms classified into three size categories. After being assured of the confidentiality of their replies, the 20 executives were asked to rate the overall performance quality of their board of directors in setting general corporate policy during the past 3-year period on a scale from 0 to 100. The scores are shown below.

Firms Classified by Size		
Large	Medium	Small
79	69	83
96	78	66
86	85	51
88	62	94
76	63	71
91	73	61
81		74

Our hypotheses would be

H_0: There is no difference in the rankings among the three groups of companies
H_1: There is a difference in the rankings among the three groups of companies

We have to enter the data in the same fashion as in the previous example. We will call the large firms group 1, the medium firms group 2, and the small firms group 3. Then, when we execute the *kruskal-wallis* command, we get the following output.

```
MTB > kruskal-wallis c1 c2

LEVEL      NOBS     MEDIAN   AVE. RANK    Z VALUE
   1          7      86.00      15.1        2.58
   2          6      71.00       7.8       -1.32
   3          7      71.00       8.1       -1.31
OVERALL     20                  10.5

H = 6.64    d.f. = 2    p = 0.037
```

MINITAB calculates the *p*-value as 0.037. This means that we will reject the null hypothesis that the rankings are the same among the different firm sizes at the 0.05 alpha level and accept the alternative that the rankings are not the same among the different size firms.

17.4 Wilcoxon Matched-Pairs Signed-Rank Test

The previous two tests have been nonparametric tests for independent samples. We will now look at a nonparametric test for two dependent samples, the Wilcoxon matched pairs signed-rank test.

Example 17.4A

Suppose the president of a company is interested in learning if there is any difference in customer satisfaction in two different stores, A and B. The ranking is from 1 to 10. A higher ranking is better. Below is a sample of 10 customers from each store.

	Customer Satisfaction	
	A	B
1	6	7
2	5	8
3	7	8
4	8	9
5	9	4
6	5	5
7	4	7
8	9	6
9	8	6
10	7	6

Our hypotheses would be

H_0: There is no difference in customer satisfaction between stores A and B.
H_1: There is a difference in customer satisfaction between stores A and B.

We will use the *wtest* command to carry out the Wilcoxon matched pairs signed-rank test.

```
MTB > wtest c3
```

Wilcoxon Signed Rank Test

```
Test of median = 0.000000 versus median not = 0.000000

            N for   Wilcoxon              Estimated
       N    Test    Statistic      P       Median
C3    10     9        21.5      0.953    0.000E+00
```

Notice that we take the difference of the two samples before we use the *wtest* command. The output shows that the *p*-value 0.953 is greater than an alpha value of either 0.10 or 0.05. Therefore, we cannot conclude that there is any difference in satisfaction between stores A and B.

Example 17.4B

We have data on net income from Lawrence, Inc., before and after a corporate reorganization. We are interested in testing to see if net income differs significantly before and after the reorganization. The data are shown below.

Market Region	Net Income Before Reorganization	Net Income After Reorganization
1	41	62
2	34	49
3	43	39
4	29	28
5	55	55
6	63	66
7	35	47
8	42	72
9	57	84
10	45	42

We will put the 'before' data in column c1 and the 'after' data in column c2.

Our hypotheses would be

H_0: There is no difference in income after reorganization.
H_1: There is a difference in income after reorganization.

The MINITAB commands and output are shown below.

```
MTB > let c3 = c2-c1
MTB > wtest c3
```

Wilcoxon Signed Rank Test

```
Test of median = 0.000000 versus median not = 0.000000
                N for   Wilcoxon                Estimated
         N      Test    Statistic       P       Median
C3       10     9       37.5            0.086   10.00
```

From the MINITAB output the calculated *p*-value for the data is 0.086. This means that at a 0.05 alpha level we cannot reject the null hypothesis that there is no difference between income before and after the reorganization.

17.5 Ranking

If we did the tests examined in this chapter manually, we would have to rank the data. For this or other reasons, we are often interested in ranking data. MINITAB has a command that can rank data in a column. But suppose we want to rank data from more than one column. In this section we will see how to do this.

Product Category	Jan-92	Jan-93
Processed poultry	198.80	192.40
Concrete ingredients	331.00	339.00
Lumber	343.00	329.60
Gas fuels	1,073.00	1,034.30
Drugs and pharmaceuticals	247.40	265.90
Synthetic fibers	157.60	151.10

Suppose we are interested in ranking the above index figures from January 1992 and January 1993. Let us assume that we had previously stored the 1992 data in column c1 and the 1993 data in column c2 on a MINITAB worksheet. The *rank* command will rank data in one column, but we want to make a single ranking of the data in both columns. We have the choice of either typing the above data again in a single column or telling MINITAB to put it in one column. Letting MINITAB put all the data in one column is called stacking the data.

The worksheet would look like the following before anything is done.

	C1	C2	C3
1	198.8	192.4	
2	331.0	339.0	
3	343.0	329.6	
4	1073.0	1034.3	
5	247.4	265.9	
6	157.6	151.1	
7			
8			

Our first step, before we can rank the above data, is to stack the data in column c1 onto c2 and put the results into c3. We will use the *stack* command to stack data.

```
MTB > stack c1 c2 c3
```

The first argument, 'c1', is the location of the first data set. The second argument, 'c2', is the location of the second data set. The third argument, 'c3', is the location where we are going to stack the two data sets. The worksheet will look like the following after stacking the two data sets.

	C1	C2	C3	C4
1	198.8	192.4	198.8	
2	331.0	339.0	331.0	
3	343.0	329.6	343.0	
4	1073.0	1034.3	1073.0	
5	247.4	265.9	247.4	
6	157.6	151.1	157.6	
7			192.4	
8			339.0	
9			329.6	
10			1034.3	
11			265.9	
12			151.1	

We can now tell MINITAB to rank the stacked data in column c3 and put the rankings in column c4.

```
MTB > rank c3 c4
```

The first argument, 'c3', is the location of the data to be ranked. The second argument, 'c4', is the location of the ranking of the data. The worksheet should look like the following after the *rank* command is issued.

	C1	C2	C3	C4	C5
1	198.8	192.4	198.8	4	
2	331.0	339.0	331.0	8	
3	343.0	329.6	343.0	10	
4	1073.0	1034.3	1073.0	12	
5	247.4	265.9	247.4	5	
6	157.6	151.1	157.6	2	
7			192.4	3	
8			339.0	9	
9			329.6	7	
10			1034.3	11	
11			265.9	6	
12			151.1	1	
13					
14					

17.6 Statistical Summary

In this chapter, we discussed three statistical tests that do not require the assumption of normality in the distribution of the population. These tests are called nonparametric statistical tests. The three tests were the Mann-Whitney U test, the Kruskal-Wallis test, and the Wilcoxon matched-pairs signed-rank test.

CHAPTER 18
TIME-SERIES:
ANALYSIS, MODEL, AND FORECASTING

18.1 Introduction

In statistics there are two kinds of data, cross-section data and time-series data. Time-series data are those recorded over time. Cross-section data pertain to a particular time. In this chapter we will look at specific issues of time-series data.

Time-series data can be broken down into four components:

1. Trend component, T
2. Cyclical component, C
3. Seasonal component, S
4. Irregular component, I

These four components can be specified as an additive model

$$x = T + C + S + I$$

or as a multiplicative model

$$x = T * C * S * I$$

Time-series data can fluctuate wildly, mostly because of these four components. In the next section we discuss how we can use the moving-average method to reduce the effects of fluctuations associated with seasonality and irregularity.

18.2 Moving Averages

We will use the quarterly earnings per share data of Johnson & Johnson (J&J) to show how moving averages can be obtained.

284   CHAPTER 18

Period, t	Earnings per Share	Period, t	Earnings per Share	Period, t	Earnings per Share	Period, t	Earnings per Share	Period, t	Earnings per Share
1	0.3	11	0.255	21	0.47	31	0.61	41	0.73
2	0.2717	12	0.21	22	0.425	32	0.475	42	1.06
3	0.2967	13	0.35	23	0.435	33	0.81		
4	0.215	14	0.37	24	0.35	34	0.785		
5	0.325	15	0.395	25	-0.375	35	0.71		
6	0.345	16	0.17	26	0.52	36	0.555		
7	0.345	17	0.39	27	0.505	37	0.95		
8	0.24	18	0.315	28	0.275	38	0.89		
9	0.415	19	0.375	29	0.68	39	0.8		
10	0.38	20	0.295	30	0.65	40	0.61		

The negative earnings in period 25 were the result of the Tylenol poisoning tragedy.

We will put the earnings per share data in column c1 of the MINITAB worksheet and name column c1 '*eps*'.

```
MTB > name c2 'eps'
MTB > SET INTO 'EPS'
DATA> .3000 .2717 .2967 .2150 .3250 .3450 .3450 .2400 .4150 .3800
DATA> .2550 .2100 .3500 .3700 .3950 .1700 .3900 .3150 .3750 .2950
DATA> .4700 .4250 .4350 .3500 -0.3750 .5200 .5050 .2750 .6800 .6500
DATA> .6100 .4750 .8100 .7850 .7100 .5550 .9500 .8900 .8000 .6100
DATA> .7300 1.0600
DATA> end
MTB > print 'eps'
```

Data Display

```
eps
    0.3000    0.2717    0.2967    0.2150    0.3250    0.3450    0.3450    0.2400
    0.4150    0.3800    0.2550    0.2100    0.3500    0.3700    0.3950    0.1700
    0.3900    0.3150    0.3750    0.2950    0.4700    0.4250    0.4350    0.3500
   -0.3750    0.5200    0.5050    0.2750    0.6800    0.6500    0.6100    0.4750
    0.8100    0.7850    0.7100    0.5550    0.9500    0.8900    0.8000    0.6100
    0.7300    1.0600
```

The formula to calculate a 4-point moving average is

$$M_t = \frac{Y_{t-2} + Y_{t-1} + Y_t + Y_{t+1}}{4}$$

where Y_t is the earnings for period t.

To follow this formula, we have to start our calculation at period 3 and end it one period before the end of the data set. The calculation for period 3 is

$$M_3 = \frac{Y_1 + Y_2 + Y_3 + Y_4}{4} = \frac{0.3 + 0.2717 + 0.2967 + 0.215}{4} = 0.2708$$

We will run the *Ma.mac* macro that comes with MINITAB. This macro is stored in the *macro* folder under the *mtbwin* folder. The first parameter of the *Ma* macro indicates the column that contains the data to be average. The second parameter of *Ma* macro indicates the moving average length.

```
MTB >   Name c3 = 'AVER1'
MTB >   %MA 'eps' 4;
SUBC>     Title "Johnson & Johnson's Moving Average Calculation";
SUBC>     Averages 'AVER1';
SUBC>     Smplot.
Executing from file: C:\PROGRAM FILES\MTBWIN\MACROS\MA.MAC
MTB >   print 'aver1'
```

Data Display

```
AVER1
       *            *            *       0.270850    0.277100    0.295425
  0.307500    0.313750    0.336250    0.345000    0.322500    0.315000
  0.298750    0.296250    0.331250    0.321250    0.331250    0.317500
  0.312500    0.343750    0.363750    0.391250    0.406250    0.420000
  0.208750    0.232500    0.250000    0.231250    0.495000    0.527500
  0.553750    0.603750    0.636250    0.670000    0.695000    0.715000
  0.750000    0.776250    0.798750    0.812500    0.757500    0.800000
```

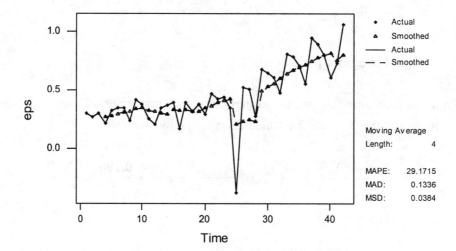

18.3 Linear Time Trend Regression

If a time-series is expected to change linearly over time, regression analysis can be used. We will use the annual sales of Ford to study the regression of a time-series.

Year	Sales (millions)	t
1968	$14,075.10	1
1969	$14,755.60	2
1970	$14,979.90	3
1971	$16,433.00	4
1972	$20,194.40	5
1973	$23,015.10	6
1974	$23,620.60	7
1975	$24,009.11	8
1976	$28,839.61	9
1977	$37,841.51	10
1978	$42,784.11	11
1979	$43,513.71	12
1980	$37,085.51	13
1981	$38,247.11	14
1982	$37,067.21	15
1983	$44,454.61	16
1984	$52,366.41	17
1985	$52,774.41	18
1986	$62,868.30	19
1987	$72,797.20	20
1988	$82,193.00	21
1989	$82,879.40	22
1990	$81,844.00	23

We will put the sales data in column c1 and name the column '*sales*'.

```
MTB > name c1 'sales'
MTB > set 'sales'
DATA> 14075.10 14755.60 14979.90 16433.00 20194.40
DATA> 23015.10 23620.60 24009.11 28839.61 37841.51
DATA> 42784.11 43513.71 37085.51 38247.11 37067.21
DATA> 44454.61 52366.41 52774.41 62868.30 72797.20
DATA> 82193.00 82879.40 81844.00
DATA> end
MTB > print 'sales'
```

Data Display

```
sales
    14075.1    14755.6    14979.9    16433.0    20194.4    23015.1    23620.6
    24009.1    28839.6    37841.5    42784.1    43513.7    37085.5    38247.1
    37067.2    44454.6    52366.4    52774.4    62868.3    72797.2    82193.0
    82879.4    81844.0
```

288   CHAPTER 18

We will run the *Trend.mac* macro that comes with MINITAB. This macro is stored in the *macro* folder under the *mtbwin* folder. The first parameter of the *Trend* macro indicates the column that contains the data to be analyzed.

```
MTB > %Trend 'sales';
SUBC>    Title "Ford Linear Time Trend".
Executing from file: C:\PROGRAM FILES\MTBWIN\MACROS\Trend.MAC
```

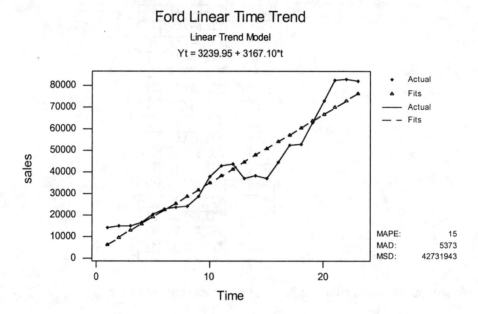

We will now regress sales over time. To do this we will put the period in column c2 and name it 'period.

```
MTB > Name c2 'period'
MTB > Set 'period'
DATA>    1( 1 : 23 / 1 )1
DATA>    End.
MTB > regress 'sales' 1 'period'
```

Regression Analysis

```
The regression equation is
sales = 3240 + 3167 period

Predictor        Coef        StDev           T          P
Constant         3240         2949        1.10      0.284
period          3167.1        215.1       14.73     0.000

S = 6841        R-Sq = 91.2%     R-Sq(adj) = 90.8%
```

Analysis of Variance

```
Source              DF           SS           MS          F        P
Regression           1  10150900144  10150900144     216.89    0.000
Residual Error      21    982834678     46801651
Total               22  11133734822
```

Unusual Observations

```
Obs    period    sales      Fit   StDev Fit   Residual   St Resid
 15      15.0    37067    50746        1566     -13679     -2.05R
```

R denotes an observation with a large standardized residual

We can see from the adjusted R-square of 90.8% that time explains sales pretty well. This indicates that Ford's sales have a linear trend. This linear trend is indicated by the coefficient of the independent variable, which is 3167. This means that the trend of Ford's sales is an increase of $3.167 billion every year.

18.4 Exponential Smoothing

Moving averages use past, current, and future data. Exponential smoothing, on the other hand, uses only current and past data. Exponential smoothing is calculated as follows:

$$s_1 = x_1$$
$$s_2 = \alpha x_2 + (1 - \alpha)s_1$$
$$s_3 = \alpha x_2 + (1 - \alpha)s_2$$
$$\ldots$$
$$s_t = \alpha x_t + (1 - \alpha)s_{t-1}$$

where α is the smoothing constant.

We will do exponential smoothing on the annual earnings per share for J&J and IBM.

J&J

YEAR	EPS
1980	1.0834
1981	1.255
1982	1.26
1983	1.285
1984	1.375
1985	1.68
1986	0.925
1987	2.415
1988	2.86
1989	3.25

IBM

YEAR	EPS
1980	6.1
1981	5.63
1982	7.39
1983	9.04
1984	10.77
1985	10.67
1986	7.81
1987	8.72
1988	9.8
1989	6.47

There is no command in MINITAB to calculate the exponential smoothing of data, so we will create a macro, *expsmth* to do this. The macro is shown below.

Expsmth Macro

```
Gmacro
noecho
# This macro is used for chapter 18
note This macro calculates exponential smoothing
note put data to calculate in column c1 of worksheet
note this macro puts exponential smoothing calculations
note in column c2 of worksheet
#define variables
name k50 = 'numtimes' #number of times to do smooth calc
name k51 = 'smconst' #smoothing constant
name k52 = 'psmconst' #previous year smooth constant
name k53 = 'period'  #period calculating
note
note How many exponential smoothing
note calculations do you want?
set 'terminal' c50;
nobs=1.
end
let 'numtimes' = c50
note What will be the smoothing constant for the current year?
set 'terminal' c50;
nobs=1.
end
let 'smconst' =c50
let 'psmconst' =1-'smconst'
note The smoothing constant for previous year will be
print 'psmconst'
note calculating ..........
#period 1 should be equal  for smoothing and regular data
let c2(1)=c1(1)

#start calculating from second period
do 'period'=2:'numtimes'+1
      let c2('period')='smconst'*c1('period')+'psmconst'*c2('period'-1)
enddo

note calculation done.
endmacro
```

We will put J&J's EPS before the exponential smoothing calculations in column c1 and name it *'eps'*. Column c2 ,*'expsmth'*, will contain the EPS after the exponential smoothing calculations. We will put the years in column c3 and name it *'year'*.

```
MTB > %expsmth
Executing from file: expsmth.MAC
This macro calculates exponential smoothing
put data to calculate in column c1 of worksheet
this macro puts exponential smoothing calculations
in column c2 of worksheet

How many exponential smoothing
calculations do you want?
DATA> 9
What will be the smoothing constant for the current year?
DATA> .3
The smoothing constant for previous year will be
```

Data Display

```
psmconst    0.700000
calculating ..........
calculation done.
MTB > print c1-c3
```

Data Display

Row	eps	expsmth	year
1	1.0834	1.08340	1980
2	1.2550	1.13488	1981
3	1.2600	1.17242	1982
4	1.2850	1.20619	1983
5	1.3750	1.25683	1984
6	1.6800	1.38378	1985
7	0.9250	1.24615	1986
8	2.4150	1.59680	1987
9	2.8600	1.97576	1988
10	3.2500	2.35803	1989

We choose nine calculations because the first exponential calculation s_1 does not contain any smoothing constant, therefore, there is no calculations. In our case of ten original data, this would leave us with only nine data to calculate. The smoothing constant is based on each individual situation. In our case we feel that a smoothing constant of 0.3 will be appropriate.

Below are the plots of the J&J EPS before and after the exponential smoothing calculations. Notice that the second plot, which contains the exponential smoothing data, is smoother than the first plot.

The plot below is before the exponential smoothing calculations.

```
MTB > Plot 'eps'*'year';
SUBC>    Connect;
SUBC>    ScFrame;
SUBC>    ScAnnotation.
```

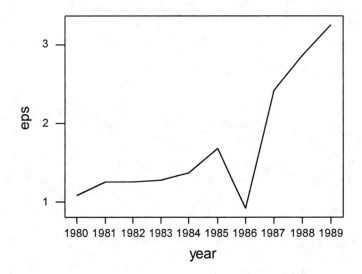

The plot below shows the effect of the exponential calculations.

```
MTB > Plot 'expsmth'*'year';
SUBC>    Connect;
SUBC>    ScFrame;
SUBC>    ScAnnotation.
```

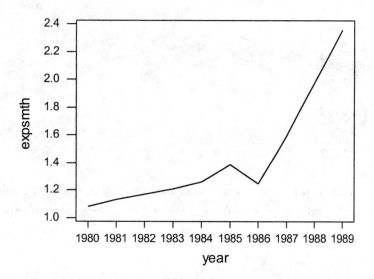

We will now illustrate the macro for the IBM data. We will put the EPS before the exponential smoothing calculations in column c1 and name it '*eps*'. Column c2, '*expsmth*', will contain the EPS after the exponential smoothing calculations. We will put the years in column c3 and name it '*year*'.

```
MTB > %expsmth
Executing from file: expsmth.MAC
This macro calculates exponential smoothing
put data to calculate in column c1 of worksheet
this macro puts exponential smoothing calculations
in column c2 of worksheet

How many exponential smoothing
calculations do you want?
DATA> 9
What will be the smoothing constant for the current year?
DATA> .3
The smoothing constant for previous year will be
```

Data Display

```
psmconst    0.700000
calculating ..........
calculation done.
MTB > print c1-c3
```

Data Display

```
Row      eps     expsmth    year
  1     6.10    6.10000    1980
  2     5.63    5.95900    1981
  3     7.39    6.38830    1982
  4     9.04    7.18381    1983
  5    10.77    8.25967    1984
  6    10.67    8.98277    1985
  7     7.81    8.63094    1986
  8     8.72    8.65766    1987
  9     9.80    9.00036    1988
 10     6.47    8.24125    1989
```

Below are plots of the IBM EPS before and after the exponential smoothing calculations. Notice that the second plot, which contains the exponential smoothing data, is smoother than the first plot.

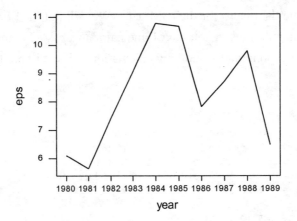

```
MTB > Plot 'expsmth'*'year';
SUBC>   Connect;
SUBC>   ScFrame;
SUBC>   ScAnnotation.
```

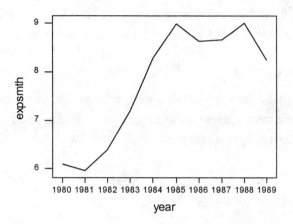

18.5 Holt-Winters

Exponential smoothing does not recognize the trend in time-series, but the Holt-Winters model does. The Holt-Winters model contains a smoothing component, s, and a trend component, T. The smoothing and trend components can be described as follows:

$$s_t = \alpha x_t + (1 - \alpha)(s_{t-1} + T_{t-1})$$
$$T_t = \beta(s_t - s_{t-1}) + (1 - \beta)T_{t-1}$$

where α and β are the exponential and trend smoothing constants respectively.

The procedure for calculating the Holt-Winters components is as follows:
1. Choose an exponential smoothing constant α between 0 and 1. A small value for α gives less weight to the current values of the time-series and more weight to the past. A large value for α gives more weight to the most recent trend of the series.
2. Choose a trend smoothing constant β between 0 and 1. Small values of β give less weight to the current changes in the level of the series and more weight to the past trend. Larger choices assign more weight to the most recent trend series.
3. Estimate the first observation of the trend T_1 by one of the following alternative methods.

METHOD 1:

Let $T_1 = 0$. If there are a large number of observations in the time-series, this method provides an adequate initial estimate of the trend.

METHOD 2:

Use the first 5 (or so) observations to estimate the initial trend by following the linear time trend regression line. Then use the estimated slope b as the first trend observation: that is, $T_1 = b$.

We will use the J&J and IBM EPS data again to calculate the Holt-Winters numbers. The data are reproduced below for convenience.

J&J

YEAR	EPS
1980	1.0834
1981	1.255
1982	1.26
1983	1.285
1984	1.375
1985	1.68
1986	0.925
1987	2.415
1988	2.86
1989	3.25

IBM

YEAR	EPS
1980	6.1
1981	5.63
1982	7.39
1983	9.04
1984	10.77
1985	10.67
1986	7.81
1987	8.72
1988	9.8
1989	6.47

We will create our own macros using the Holt-Winters calculation. The three macros are called, *holtzero, holtreg,* and *holtb*. We use the *holtzero* macro if we want to use method 1 in calculating the first observation. We use the *holtreg* macro if we want to use method 2 in calculating the first observation. The two macros are shown below.

Holtzero Macro

```
gmacro
noecho
# This macro is used for chapter 18
note This macro calculates the holt-winters calculations
note put data to calculate in column c1 of worksheet
note this macro puts the holt-winters calculations
note in columns c2 and c3 of the worksheet

#Define variables
name k50 = 'times'
name k51 = 'expcnst'
name k52 = 'trndcnst'
name k53 = 'period'

let c2(1)=c1(1) #perod 1 smooth equals data
let c3(1)=0 #period 1 trend = 0

note
note How many holt-winters
note calculations do you want?
set 'terminal' c50;
nobs=1.
end
let 'times'=c50
erase c50

note What is the exponential smoothing constant?
set 'terminal' c50;
nobs=1.
End
let 'expcnst'=c50
erase c50

note What is the trend smoothing constant?
set 'terminal' c50;
nobs=1.
end
let 'trndcnst'=c50
erase c50

note calculating ..........

#this macro is referenced by holtzero and holtreg macro
#this macro is used in chapter 18
do 'period' = 2:'times'+1

 let c2('period')='expcnst'*c1('period')+(1-'expcnst')*(c2('period'-1)+c3('period'-1))
 let c3('period')='trndcnst'*(c2('period')-c2('period'-1))+(1-'trndcnst')*c3('period'-1)
enddo

name c1='eps' c2='smooth' c3='trend'
print c1-c3
endmacro
```

Holtreg Macro

```
gmacro
noecho
# This macro is used for chapter 18
note This macro calculates the holt-winters calculations
note put data to calculate in column c1 of worksheet
note this macro puts the holt-winters calculations
note in columns c2 and c3 of the worksheet

#Define variables
name k50 = 'times'
name k51 = 'expcnst'
name k52 = 'trndcnst'
name k53 = 'period'

let c2(1)=c1(1)
let c50(1)=c1(1)
let c50(2)=c1(2)
let c50(3)=c1(3)
let c50(4)=c1(4)
let c50(5)=c1(5)
let c51(1)=1
let c51(2)=2
let c51(3)=3
let c51(4)=4
let c51(5)=5
regress c50 1 c51;
coefficient c52.
let c3(1)=c52(2)
erase c50-c52

note
note How many holt-winters
note calculations do you want?
set 'terminal' c50;
nobs=1.
end
let 'times'=c50
erase c50

note What is the exponential smoothing constant?
set 'terminal' c50;
nobs=1.
End
let 'expcnst'=c50
erase c50

note What is the trend smoothing constant?
set 'terminal' c50;
nobs=1.
end
let 'trndcnst'=c50
erase c50

note calculating ..........

do 'period' = 2:'times'+1
  let c2('period')='expcnst'*c1('period')+(1-'expcnst')*(c2('period'-1)+c3('period'-1))
  let c3('period')='trndcnst'*(c2('period')-c2('period'-1))+(1-'trndcnst')*c3('period'-1)
enddo

name c1='eps' c2='smooth' c3='trend'
print c1-c3
endmacro
```

We will use the J&J EPS data to illustrate method 1, the *holtzero* macro. As illustrated below, we begin by putting the J&J EPS data into column c1 of the worksheet.

```
MTB > %holtzero
Executing from file: holtzero.MAC
This macro calculates the holt-winters calculations
put data to calculate in column c1 of worksheet
this macro puts the holt-winters calculations
in columns c2 and c3 of the worksheet

How many holt-winters
calculations do you want?
DATA> 9
What is the exponential smoothing constant?
DATA> .3
What is the trend smoothing constant?
DATA> .2
calculating ..........
```

Data Display

Row	eps	smooth	trend
1	1.0834	1.08340	0.000000
2	1.2550	1.13488	0.010296
3	1.2600	1.17962	0.017185
4	1.2850	1.22327	0.022477
5	1.3750	1.28452	0.030232
6	1.6800	1.42433	0.052147
7	0.9250	1.31103	0.019059
8	2.4150	1.65556	0.084153
9	2.8600	2.07580	0.151370
10	3.2500	2.53402	0.212740

We will now plot the smoothed and trend data in separate plots and then plot the smoothed, trend, and original EPS data together in one graph. To get a meaningful horizontal axis, we will also put the years in column c4 and name that column '*year*'.

```
MTB > Plot 'smooth'*'year';
SUBC>   Connect;
SUBC>   Minimum 2 0;
SUBC>   Maximum 2 3.6;
SUBC>   ScFrame;
SUBC>   ScAnnotation;
SUBC>   Axis 1;
SUBC>   Axis 2.
```

300   CHAPTER 18

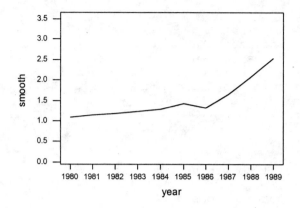

```
MTB > Plot 'trend'*'year';
SUBC>   Connect;
SUBC>   Minimum 2 0;
SUBC>   Maximum 2 3.6;
SUBC>   ScFrame;
SUBC>   ScAnnotation;
SUBC>   Axis 1;
SUBC>   Axis 2.
```

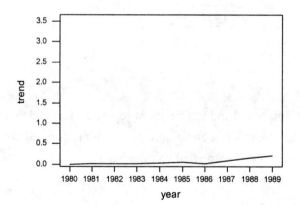

```
MTB > Plot 'trend'*'year' 'smooth'*'year' 'eps'*'year';
SUBC>    Connect;
SUBC>    Overlay;
SUBC>    Minimum 2 0;
SUBC>    Maximum 2 3.6;
SUBC>    ScFrame;
SUBC>    ScAnnotation;
SUBC>    Axis 1;
SUBC>    Axis 2;
SUBC>       Label "trend,smooth,eps".
```

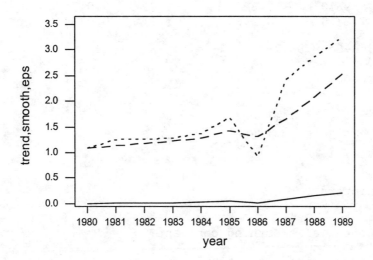

To illustrate method 2 and the *holtreg* macro, we will use the IBM EPS data. This is illustrated below. The IBM EPS data are put into column c1 of the worksheet, and the years into column c4.

```
MTB > %holtreg
Executing from file: holtreg.MAC
This macro calculates the holt-winters calculations
put data to calculate in column c1 of worksheet
this macro puts the holt-winters calculations
in columns c2 and c3 of the worksheet
```

Regression Analysis

The regression equation is
C50 = 3.96 + 1.27 C51

Predictor	Coef	StDev	T	P
Constant	3.9610	0.8277	4.79	0.017
C51	1.2750	0.2496	5.11	0.015

S = 0.7891   R-Sq = 89.7%   R-Sq(adj) = 86.3%

Analysis of Variance

Source	DF	SS	MS	F	P
Regression	1	16.256	16.256	26.10	0.015
Residual Error	3	1.868	0.623		
Total	4	18.125			

```
How many holt-winters
calculations do you want?
DATA> 9
What is the exponential smoothing constant?
DATA> .3
What is the trend smoothing constant?
DATA> .2
calculating ...........
```

Data Display

Row	eps	smooth	trend
1	6.10	6.1000	1.27500
2	5.63	6.8515	1.17030
3	7.39	7.8323	1.13239
4	9.04	8.9873	1.13691
5	10.77	10.3179	1.17566
6	10.67	11.2465	1.12625
7	7.81	11.0039	0.85248
8	8.72	10.9155	0.66430
9	9.80	11.0458	0.55751
10	6.47	10.0634	0.24951

We will now plot the smoothed and trend data in separate plots and then put the smoothing, trend, and EPS data together in the same plot.

```
MTB > Plot 'smooth'*'year';
SUBC>   Connect;
SUBC>   Minimum 2 0;
SUBC>   Maximum 2 12;
SUBC>   ScFrame;
SUBC>   ScAnnotation.
```

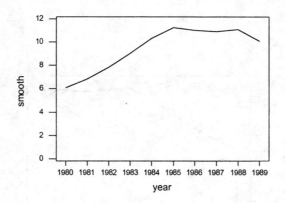

```
MTB > Plot 'trend'*'year';
SUBC>   Connect;
SUBC>   Minimum 2 0;
SUBC>   Maximum 2 12;
SUBC>   ScFrame;
SUBC>   ScAnnotation.
```

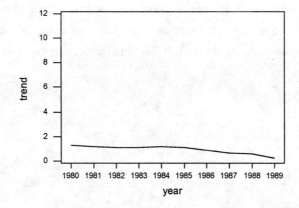

```
MTB > Plot 'smooth'*'year' 'trend'*'year' 'eps'*'year';
SUBC>    Connect;
SUBC>    Overlay;
SUBC>    Minimum 2 0;
SUBC>    Maximum 2 12;
SUBC>    ScFrame;
SUBC>    ScAnnotation;
SUBC>    Axis 1;
SUBC>    Axis 2;
SUBC>       Label "smooth,trend,eps".
```

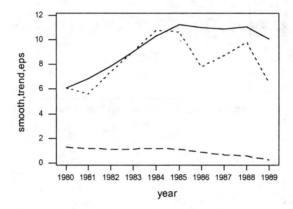

18.6 Statistical Summary

In this chapter, we first discuss how the moving-average method can be used to remove both seasonal and irregular components of time-series data. Then we discussed how linear time trend regression, exponential smoothing, and the Holt-Winters model can be used to analyze the time-series data.

CHAPTER 19
INDEX NUMBERS AND
STOCK MARKET INDEXES

19.1 Introduction

In this chapter we will look at indexes. Index numbers are numbers that compare an activity in one time or place to a similar activity in a specific base period. The first index that we will look at is the simple price index of a single item. When used this way we can measure the relative price change over time.

19.2 Simple Price Index

The formula for the simple price index is

$$I_t = \frac{P_t}{P_0} * 100$$

where P_t is the price in period t, P_0 is the base period price, and I_t is the simple price index at time t.

Example 19.2A

Suppose we would like to create a simple price index for eggs. We know the price of eggs for the following years.

PRICE OF EGGS

YEAR	PRICE
1991	$1.00
1992	$1.20
1993	$1.50

In creating an index we have to decide on a base year. Let us choose 1991 as the base year in this case. To use MINITAB to create the index, we should put the years in column c1 and the prices in column c2. The worksheet, when finished, should look like the following.

	C1	C2	C3
	year	price	index
1	1991	1.0	
2	1992	1.2	
3	1993	1.5	
4			
5			
6			
7			

The index numbers will go in column c3. The index numbers can be easily created by using the *let* command as illustrated below.

```
MTB > let c3=c2/c2(1) * 100
MTB > print c1-c3
```

Data Display

```
Row    year   price   index

  1    1991    1.0     100
  2    1992    1.2     120
  3    1993    1.5     150
```

In the above expression, c2 is the numerator of the simple price index and c2(1) is the denominator or the simple price index.

The index numbers tell us that the price of eggs in 1992 was 20% greater than it was in 1991, or 120% of the price of eggs in 1991. The index numbers also tell us that the price of eggs in 1993 was 50% greater than it was in 1991, or 150% of the price of eggs in 1991.

We will create custom functions in Excel to calculate the indexes in this chapter. The function to calculate the simple price index is shown below.

```
'/****************************************************************************
'/Purpose: Calculates the price index given the current year and base year
'/Parameter: CurrentYear: Range that contains current year.  Must be only one cell
'/          BaseYear: Range that contains the base year.  Must be only one cell
'/****************************************************************************
Function PriceIndex(CurrentYear As Range, BaseYear As Range) As Double
        'checks to see if current year range is greater than 1 cell
        If CurrentYear.Cells.Count > 1 Then
            Call MsgBox("In cell " & ActiveCell.Address & _
            " please choose only one cell for the current year", vbCritical)
        'checks to see if base year range is greater than 1 cell
        ElseIf BaseYear.Cells.Count > 1 Then
            Call MsgBox("In cell " & ActiveCell.Address & _
            " please choose only one cell for the current year", vbCritical)
        Else
            'No errors.  Calcualte the Price Index.
            PriceIndex = (CurrentYear / BaseYear) * 100
        End If
End Function
```

In Excel input the price of eggs data and enter the PriceIndex function as shown below.

	A	B	C
1	Year	Price	Index
2	1991	1	=PriceIndex(B2,B2)
3	1992	1.2	=PriceIndex(B3,B2)
4	1993	1.5	=PriceIndex(B4,B2)
5			

After entering the PriceIndex function, your spreadsheet should look like the following.

	A	B	C	D
1	Year	Price	Index	
2	1991	1	100	
3	1992	1.2	120	
4	1993	1.5	150	

In studying index numbers it is important to remember what the base period is. Otherwise, we may end up with misinformation.

Example 19.2B

Instead of choosing 1991 as the base year for indexing the eggs, let us choose 1992 as the base year.

```
MTB > let c3=c2/c2(2) * 100
MTB > print c1-c3
```

Data Display

```
Row     year    price      index

  1     1991    1.0        83.333
  2     1992    1.2       100.000
  3     1993    1.5       125.000
```

We can see that after choosing a different base year the index number for every year is different. The index for the year 1991 was 100 when the base year was 1991, but now it is 83.333. This example also shows that an index number can be less than 100, indicating that the price (or other variable) is less than in the base year.

Instead of having an index for a single item, we can have an index for a group of items. The mathematical formula for a group of items is shown below.

$$I_t = \frac{\sum_{i=1}^{n} P_{ti}}{\sum_{i=1}^{n} P_{0i}} * 100$$

where P_{ti} is the price at period t for item i, P_{0i} is the base period price for item i, and I_t is the simple price index at time t.

The consumer price index (CPI) is an example of a simple index for a group of items. The CPI has around 2000 items but we will use only four:

      one dozen eggs
      one gallon of milk
      one pound of butter
      one loaf of bread

The data for the four items are

Year	Eggs	Milk	Butter	Bread
1991	1.00	1.50	1.10	0.40
1992	1.20	1.75	1.35	0.70
1993	1.50	2.00	1.60	0.90

The MINITAB worksheet should look like the following after the data are entered.

	C1 year	C2 eggs	C3 milk	C4 butter	C5 bread	C6 index
1	1991	1.0	1.50	1.10	0.4	
2	1992	1.2	1.75	1.35	0.7	
3	1993	1.5	2.00	1.60	0.9	

We will again use 1991 as the base year and create the index numbers with only the *let* command. This is illustrated below.

```
MTB > let c6=((c2+c3+c4+c5)/(c2(1)+c3(1)+C4(1)+c5(1)))*100
MTB > print c1-c6
```

Data Display

```
Row   year   eggs   milk   butter   bread   index

 1    1991   1.0    1.50   1.10     0.4     100
 2    1992   1.2    1.75   1.35     0.7     125
 3    1993   1.5    2.00   1.60     0.9     150
```

In the *let* expression (c2+c3+c4+c5) represents the numerator of the simple index of a group of items. The (c2(1)+c3(1)+C4(1)+c5(1)) represents the denominator of the simple index of a group of items.

The MINITAB output tells us that the 1992 group of goods is 25% more expensive than the same group of goods in 1991, or 125% of the price of the same group in 1991. The MINITAB output also tells us that the group of goods in 1993 is 50% more expensive than the same group in 1991, or 150% of the price of the goods in 1991.

CHAPTER 19

19.3 LASPEYRES Price Index

Previously we have only looked at price in our index. The Laspeyres price index takes into consideration both price and quantity. The base period quantity is in both the numerator and the denominator of the Laspeyres index. The formula for the Laspeyres index is

$$I_t = \frac{\sum_{i=1}^{n} P_{ti} Q_{0i}}{\sum_{i=1}^{n} P_{0i} Q_{0i}} * 100$$

where P_{0i} is the base price of item i, Q_{0i} is the base quantity of item i, P_{ti} is the price at time t of item i, and I_t is the Laspeyres price index.

We will use the following items to calculate the Laspeyres price index.

	Eggs		Milk		Butter		Bread		Shirts	
Year	Price	Quantity	Price	Quantity	Price	Quantity	Price	Quantity	Price	Quantity
1991	1.00	150	1.50	300	1.10	200.00	0.40	1100	16.00	10
1992	1.20	160	1.75	250	1.35	180.00	0.70	1000	20.00	15
1993	1.50	180	2.00	300	1.60	250.00	0.90	1050	10.00	20

The MINITAB worksheet should look like the following after inputting the data.

	C1	C2	C3	C4	C5-T	C6	C7
↓	quantity	1991	1992	1993	commod	indexyr	index
1	150	1.0	1.20	1.5	eggs	1991	
2	300	1.5	1.75	2.0	milk	1992	
3	200	1.1	1.35	1.6	butter	1993	
4	1100	0.4	0.70	0.9	bread		
5	10	16.0	20.00	10.0	shirts		

Note that the quantity of the base year, 1991, is entered in column c1. We then put the prices for each year in a separate column. Column c5 contains descriptions of the content, of each row. The first row is the data for eggs, the second for milk, the third for butter, the fourth bread, the fifth for shirts. The '*indexyr*' column is where we put the index for each year.

Note that in the worksheet, only the quantity for the base year is entered because the Laspeyres price index only has the base year quantity, Q_{0i}, in its formula.

The MINITAB commands for the Laspeyres price index are illustrated below.

```
MTB > let c7(1)=(sum(c1*c2)/sum(c1*c2)) * 100
MTB > let c7(2)=(sum(c1*c3)/sum(c1*c2)) * 100
MTB > let c7(3)=(sum(c1*c4)/sum(c1*c2)) * 100
MTB > print c1-c7
```

Data Display

Row	quantity	1991	1992	1993	commod	indexyr	index
1	150	1.0	1.20	1.5	eggs	1991	100.000
2	300	1.5	1.75	2.0	milk	1992	136.972
3	200	1.1	1.35	1.6	butter	1993	157.394
4	1100	0.4	0.70	0.9	bread		
5	10	16.0	20.00	10.0	shirts		

From the MINITAB output we can see that the 1992 basket of goods is 37% more expensive than the 1991 basket of goods and that the 1993 basket of goods is 57% more expensive than the 1991 basket of goods.

The following exhibit shows the various parts of the *let* command expression that correspond to the Laspeyres price index.

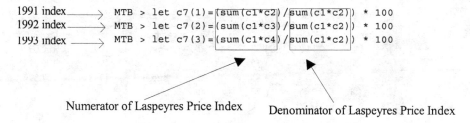

We will now create a custom Laspeyres Price Index function in Excel. The Excel function code to calculate the Laspeyres Price Index is shown below.

```
'/*******************************************************************
'/Purpose: Calculates the Laspeyres price index  given the current year and base year
'/         and the quantity of the base year.
'/Parameter: CurrentYear: Range that contains current year.
'/           BaseYear: Range that contains the base year.
'/           BaseQuantity: Range that contains the base year quantity of the base year.
'/*******************************************************************
Function LaspeyresPriceIndex(BaseQuantity As Range, BaseYear As Range, _
                      CurrentYear As Range) As Double
        'the sumproduct is a worksheet function
        LaspeyresPriceIndex = (WorksheetFunction.SumProduct(BaseQuantity, CurrentYear)/ _
                      WorksheetFunction.SumProduct(BaseQuantity, BaseYear)) _
                      * 100

End Function
```

Enter the data as shown below in Excel.

	A	B	C	D	E	F	G
10	quantity	1991	1992	1993	commod	indexyr	index
11	150	1.00	1.20	1.50	eggs	1991	
12	300	1.50	1.75	2.00	milk	1992	
13	200	1.10	1.35	1.60	butter	1993	
14	1100	0.40	0.70	0.90	bread		
15	10	16.00	20.00	10.00	shirts		

In column G, enter the LaspeyresPriceIndex function as shown below.

	G
10	index
11	=LaspeyresPriceIndex(A11:A15,B11:B15,B11:B15)
12	=LaspeyresPriceIndex(A11:A15,B11:B15,C11:C15)
13	=LaspeyresPriceIndex(A11:A15,B11:B15,D11:D15)

After entering the data and function properly your Excel worksheet should look like the following.

	A	B	C	D	E	F	G
10	quantity	1991	1992	1993	commod	indexyr	index
11	150	1.00	1.20	1.50	eggs	1991	100.000
12	300	1.50	1.75	2.00	milk	1992	136.972
13	200	1.10	1.35	1.60	butter	1993	157.394
14	1100	0.40	0.70	0.90	bread		
15	10	16.00	20.00	10.00	shirts		

19.4 PAASCHE Price Index

The Paasche price index is like the Laspeyres index except that it uses the current-year quantities instead of the base-year quantity. The formula for the Paasche price index is

$$I_t = \frac{\sum_{i=1}^{n} P_{ti} Q_{ti}}{\sum_{i=1}^{n} P_{0i} Q_{ti}} * 100$$

We will use the same data that we did for the Laspeyres price index. It is reproduced below for convenience.

	Eggs		Milk		Butter		Bread		Shirts	
Year	Price	Quantity	Price	Quantity	Price	Quantity	Price	Quantity	Price	Quantity
1991	1.00	150	1.50	300	1.10	200.00	0.40	1100	16.00	10
1992	1.20	160	1.75	250	1.35	180.00	0.70	1000	20.00	15
1993	1.50	180	2.00	300	1.60	250.00	0.90	1050	10.00	20

The MINITAB worksheet should look like the following after inputting the data.

	C1	C2	C3	C4	C5	C6	C7-T	C8	C9
	quant91	1991	quant92	1992	quant93	1993	commod	indexyr	index
1	150	1.0	160	1.20	180	1.5	eggs	1991	
2	300	1.5	250	1.75	300	2.0	milk	1992	
3	200	1.1	180	1.35	250	1.6	butter	1993	
4	1100	0.4	1000	0.70	1050	0.9	bread		
5	10	16.0	15	20.00	20	10.0	shirts		

```
MTB > let c9(1)=(sum(c1*c2)/sum(c1*c2)) * 100
MTB > let c9(2)=(sum(c3*c4)/sum(c3*c2)) * 100
MTB > let c9(3)=(sum(c5*c6)/sum(c5*c2)) * 100
MTB > print c1-c9
```

Data Display

```
Row   quant91    1991   quant92    1992   quant93    1993   commod   indexyr

 1        150     1.0       160    1.20       180     1.5   eggs        1991
 2        300     1.5       250    1.75       300     2.0   milk        1992
 3        200     1.1       180    1.35       250     1.6   butter      1993
 4       1100     0.4      1000    0.70      1050     0.9   bread
 5         10    16.0        15   20.00        20    10.0   shirts

index
  100.000    136.380    146.809
```

From the computer output we can see that the 1992 price of goods is 37% greater than the price of goods in 1991. We can see that the 1993 price of goods is 47% greater than the price of goods in 1991.

The following exhibit shows the various parts of the *let* command expression that correspond to the Paasche price index.

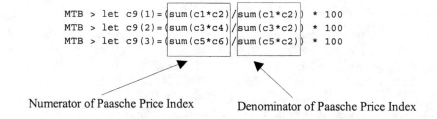

We will now create a custom Paasche price index in Excel. The Excel function code to calculate the Paasche price index is shown below.

```
'/****************************************************************************
'/Purpose: Calculates the Paasche price index  given the current year and base year
'/         and the current quantity
'/Parameter: CurrentYear: Range that contains current year.
'/           BaseYear: Range that contains the base year.
'/           CurrentQuantity: Range that contains the current year quantity.
'/****************************************************************************
Function PaaschePriceIndex(CurrentQuantity As Range, CurrentYear As Range, _
                     BaseYear As Range) As Double


     PaaschePriceIndex = (WorksheetFunction.SumProduct(CurrentQuantity, CurrentYear)/_
                   WorksheetFunction.SumProduct(CurrentQuantity, BaseYear)) _
                   * 100

End Function
```

Enter the data in Excel as shown below.

	A	B	C	D	E	F	G	H	I
20	quant91	1991	quant92	1992	quant93	1993	commod	indexyr	index
21	150	1.00	160	1.20	180	1.50	eggs	1991	
22	300	1.50	250	1.75	300	2.00	milk	1992	
23	200	1.10	180	1.35	250	1.60	butter	1993	
24	1100	0.40	1000	0.70	1050	0.90	bread		
25	10	16.00	15	20.00	20	10.00	shirts		
26									

INDEX NUMBERS AND STOCK MARKET INDEXES 315

Enter the Paasche price index function as shown.

	I
20	index
21	=PaaschePriceIndex(A21:A25,B21:B25,B21:B25)
22	=PaaschePriceIndex(C21:C25,D21:D25,B21:B25)
23	=PaaschePriceIndex(E21:E25,F21:F25,B21:B25)

After entering the Paasche price index function the Excel worksheet should look like the following.

	A	B	C	D	E	F	G	H	I
20	quant91	1991	quant92	1992	quant93	1993	commod	indexyr	index
21	150	1.00	160	1.20	180	1.50	eggs	1991	100.000
22	300	1.50	250	1.75	300	2.00	milk	1992	136.380
23	200	1.10	180	1.35	250	1.60	butter	1993	146.809
24	1100	0.40	1000	0.70	1050	0.90	bread		
25	10	16.00	15	20.00	20	10.00	shirts		

19.5 Fisher's Ideal Price Index

Fisher's ideal price index, which contains both the Laspeyres price index and the Paasche price index, has a value between those two indexes. The mathematical formula for the Fisher's ideal price index is

$$FI = \sqrt{\frac{\sum_{i=1}^{n} P_{ti}Q_{0i}}{\sum_{i=1}^{n} P_{0i}Q_{0i}} * \frac{\sum_{i=1}^{n} P_{ti}Q_{ti}}{\sum_{i=1}^{n} P_{0i}Q_{ti}}}$$

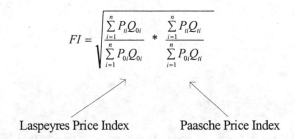

      Laspeyres Price Index   Paasche Price Index

We will use the same data as the Laspeyres price index. It is reproduced below for convenience.

	Eggs		Milk		Butter		Bread		Shirts	
Year	Price	Quantity	Price	Quantity	Price	Quantity	Price	Quantity	Price	Quantity
1991	1.00	150	1.50	300	1.10	200.00	0.40	1100	16.00	10
1992	1.20	160	1.75	250	1.35	180.00	0.70	1000	20.00	15
1993	1.50	180	2.00	300	1.60	250.00	0.90	1050	10.00	20

The MINITAB worksheet should look like the following after inputting the data.

	C1	C2	C3	C4	C5	C6	C7-T	C8	C9
	quant91	1991	quant92	1992	quant93	1993	commod	indexyr	index
1	150	1.0	160	1.20	180	1.5	eggs	1991	
2	300	1.5	250	1.75	300	2.0	milk	1992	
3	200	1.1	180	1.35	250	1.6	butter	1993	
4	1100	0.4	1000	0.70	1050	0.9	bread		
5	10	16.0	15	20.00	20	10.0	shirts		

```
MTB > let c9(1)=sqrt((sum(c1*c2)/sum(c1*c2))*(sum(c1*c2)/sum(c1*c2))) * 100
MTB > let c9(2)=sqrt((sum(c1*c4)/sum(c1*c2))*(sum(c3*c4)/sum(c3*c2))) * 100
MTB > let c9(3)=sqrt((sum(c1*c6)/sum(c1*c2))*(sum(c5*c6)/sum(c5*c2))) * 100
MTB > print c1-c9
```

Data Display

```
Row   quant91    1991   quant92    1992   quant93    1993   commod    indexyr

  1       150     1.0       160    1.20       180     1.5   eggs         1991
  2       300     1.5       250    1.75       300     2.0   milk         1992
  3       200     1.1       180    1.35       250     1.6   butter       1993
  4      1100     0.4      1000    0.70      1050     0.9   bread
  5        10    16.0        15   20.00        20    10.0   shirts

index
  100.000    136.676    152.009
```

From the MINITAB output we can see that the price of goods in 1992 is 36% greater than the price of goods in 1991. We can also see that the price of goods in 1993 is 52% greater than the price of goods in 1991.

The following exhibit shows the various parts of the *let* command expression that correspond to the Fisher's ideal price index.

INDEX NUMBERS AND STOCK MARKET INDEXES

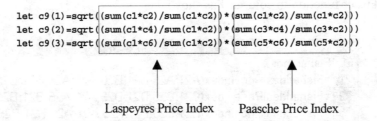

We will now create a custom Fisher's ideal price index in Excel. The Excel function code to calculate the Fisher's ideal price index is shown below.

```
'/************************************************************************
'/Purpose: Calculates the Fisher Ideal Price Index. Calculates the index by
'          calling the Laseyres Price Index and Paasche Price Index.
'/Parameter: CurrentYear: Range that contains current year.
'/           BaseYear: Range that contains the base year.
'/           BaseQuantity: Range that contains the base year quantity of the base year.
'/           CurrentQuantity: Range that contains the current year quantity
'/************************************************************************
Function FisherIdealPriceIndex(CurrentQuantity As Range, CurrentYear As Range, _
                    BaseQuantity As Range, BaseYear As Range) As Double

    FisherIdealPriceIndex = Sqr(LaspeyresPriceIndex(BaseQuantity, BaseYear, CurrentYear) _
                    * PaaschePriceIndex(CurrentQuantity, CurrentYear, BaseYear))

End Function
```

To illustrate the Fisher Ideal price index we will use the Paasche price index data that we entered in Excel. We will put the Fisher Index in column J as shown below.

	A	B	C	D	E	F	G	H	I	J
20	quant91	1991	quant92	1992	quant93	1993	commod	indexyr	index	Fisher Index
21	150	1.00	160	1.20	180	1.50	eggs	1991	100.000	
22	300	1.50	250	1.75	300	2.00	milk	1992	136.380	
23	200	1.10	180	1.35	250	1.60	butter	1993	146.809	
24	1100	0.40	1000	0.70	1050	0.90	bread			
25	10	16.00	15	20.00	20	10.00	shirts			

Enter the Fisher function in column J as shown below.

	J
20	Fisher Index
21	=FisherIdealPriceIndex(A21:A25,B21:B25,A21:A25,B21:B25)
22	=FisherIdealPriceIndex(C21:C25,D21:D25,A21:A25,B21:B25)
23	=FisherIdealPriceIndex(E21:E25,F21:F25,A21:A25,B21:B25)

The Excel worksheet should look like the following after inputting the Fisher Index.

	A	B	C	D	E	F	G	H	I	J
20	quant91	1991	quant92	1992	quant93	1993	commod	indexyr	index	Fisher Index
21	150	1.00	160	1.20	180	1.50	eggs	1991	100.000	100.000
22	300	1.50	250	1.75	300	2.00	milk	1992	136.380	136.676
23	200	1.10	180	1.35	250	1.60	butter	1993	146.809	152.009
24	1100	0.40	1000	0.70	1050	0.90	bread			
25	10	16.00	15	20.00	20	10.00	shirts			

19.6 Laspeyres Quantity Index

The Laspeyres quantity index is very much like the Laspeyres price index. The difference between the two is that the Laspeyres quantity index keeps the price in the numerator constant while the Laspeyres price index keeps the quantity in the numerator constant. The formula for the Laspeyres quantity index is shown below.

$$I_t = \frac{\sum_{i=1}^{n} P_{0i} Q_{ti}}{\sum_{i=1}^{n} P_{0i} Q_{0i}} * 100$$

The Laspeyres quantity index represents the total cost of the quantities in the year in question at base-year prices as a percentage of the total cost of the base-year quantities. Because prices are kept constant, any change in the index is due to the change in quantities between the base year and the year in question.

We will use the following data to illustrate the Laspeyres quantity index.

Year	1991 Price	1991 Quantity	199 Quantit	1993 Quantity
Automobiles	1000	40	50	60
Computers	500	30	40	50
Televisions	200	10	20	30

The MINITAB worksheet should look like the following.

	C1	C2	C3	C4	C5-T	C6	C7
↓	quantity	1991	1992	1993	command	indexyr	index
1	1000	40	50	60	auto	1991	
2	500	30	40	50	computer	1992	
3	200	10	20	30	tele	1993	
4							

```
MTB > let c7(1)=(sum(c1*c2)/sum(c1*c2)) * 100
MTB > let c7(2)=(sum(c1*c3)/sum(c1*c2)) * 100
MTB > let c7(3)=(sum(c1*c4)/sum(c1*c2)) * 100
MTB > print c1-c7

Data Display

Row   quantity    1991    1992    1993    command    indexyr      index

  1       1000      40      50      60    auto          1991    100.000
  2        500      30      40      50    computer      1992    129.825
  3        200      10      20      30    tele          1993    159.649
```

Notice that the commands for the Laspeyres quantity index are the same as for the Laspeyres price index. This is possible because the worksheet format is the same for both.

In the MINITAB output, the indexes for 1992 and 1993 indicate that the cost of the three commodities increased 30% and 60%, respectively. The price has been held constant, so changes in the indexes are due to changes in the quantities of the commodities for each period. In other words, the 1992 and 1993 indexes show that the quantities of the goods increased 30% and 60%, respectively, from the 1991 base year.

320   CHAPTER 19

We will now create a custom Laspeyres Quantity Index function in Excel. The Excel function code to calculate the Laspeyres Price Index is shown below.

```
'/*******************************************************************************
'/Purpose: Calculates the Laspeyres price index  given the current year and base year
'/         and the quantity of the base year.
'/Parameter: CurrentQuantity: Range that contains current year price.
'/           BasePrice: Range that contains the base year price.
'/           BaseQuantity: Range that contains the base year quantity of the base year.
'/*******************************************************************************
Function LaspeyresQuantityIndex(BaseQuantity As Range, BasePrice As Range, _
                       CurrentQuantity As Range) As Double

    LaspeyresQuantityIndex = (WorksheetFunction.SumProduct(BasePrice, CurrentQuantity) / _
                    WorksheetFunction.SumProduct(BaseQuantity, BasePrice)) _
                    * 100

End Function
```

To illustrate the Laspeyres quantity index, enter data in Excel as shown below.

	A	B	C	D	E	F	G
30	Price	1991	1992	1993	commod	indexyr	index
31	1000	40	50	60	auto	1991	
32	500	30	40	50	computer	1992	
33	200	10	20	30	tele	1993	

In column G, enter the Laspeyres quantity index function as shown below.

	G
30	index
31	=LaspeyresQuantityIndex(B31:B33,B31:B33,A31:A33)
32	=LaspeyresQuantityIndex(B31:B33,C31:C33,A31:A33)
33	=LaspeyresQuantityIndex(B31:B33,D31:D33,A31:A33)

After entering the Laspeyres quantity index function the Excel data sheet should look like the following.

	A	B	C	D	E	F	G
30	Price	1991	1992	1993	commod	indexyr	index
31	1000	40	50	60	auto	1991	100.000
32	500	30	40	50	computer	1992	129.825
33	200	10	20	30	tele	1993	159.649

19.7 Paache Quantity Index

The Paasche quantity index is like the Paasche price index. The only difference is that the Paasche price index denominator holds the price constant, while the Paasche quantity index holds the quantity constant in the denominator. The formula for the Paasche quantity index is

$$I_t = \frac{\sum_{i=1}^{n} P_{ti} Q_{ti}}{\sum_{i=1}^{n} P_{ti} Q_{0i}} * 100$$

The MINITAB worksheet should look like the following after inputting the data.

	C1	C2	C3	C4	C5	C6	C7-T	C8	C9
	1991	quant91	1992	quant92	1993	quant93	commod	indexyr	index
1	2000	40	2100	50	2500	60	auto	1991	
2	600	30	600	40	700	50	comput	1992	
3	250	10	300	20	300	20	tele	1993	
4									

```
MTB > let c9(1)=sum(c1*c2)/sum(c1*c2) * 100
MTB > let c9(2)=sum(c3*c4)/sum(c3*c2) * 100
MTB > let c9(3)=sum(c5*c6)/sum(c5*c2) * 100
MTB > print c1-c9
```

Data Display

```
Row     1991   quant91      1992   quant92      1993   quant93   commod   indexyr

 1      2000        40      2100        50      2500        60   auto        1991
 2       600        30       600        40       700        50   comput      1992
 3       250        10       300        20       300        20   tele        1993

index
 100.000    128.571    154.032
```

Prices are held fixed in the equation, so a change between numerator and denominator reflects a change in the quantities over three years. In other words, 1992 and 1993 saw a 29% and 54% increase in quantity, respectively, over the 1991 quantity.

We will now create a custom Paasche quantity index function in Excel. The Excel function code to calculate the Paasche quantity index is shown below.

```
'/**************************************************************************
'/Purpose: Calculates the Paasche quantity index  given the current price and base
'/          quantity and the current quantity
'/Parameter: CurrentPrice: Range that contains current year price.
'/           BaseQuantity: Range that contains the base quantity.
'/           CurrentQuantity: Range that contains the current year quantity.
'/**************************************************************************
Function PaascheQuantityIndex(CurrentQuantity As Range, CurrentPrice As Range, _
                    BaseQuantity As Range) As Double


  PaascheQuantityIndex = (WorksheetFunction.SumProduct(CurrentQuantity, CurrentPrice) / _
                WorksheetFunction.SumProduct(BaseQuantity, CurrentPrice)) _
                * 100

End Function
```

To illustrate the Paasche quantity index, enter data in Excel as shown below.

	A	B	C	D	E	F	G	H	I
40	1991	quant91	1992	quant92	1993	quant93	commod	indexyr	index
41	2000	40	2100	50	2500	60	auto	1991	
42	600	30	600	40	700	50	comput	1992	
43	250	10	300	20	300	20	tele	1993	

In column I, enter the Paasche quantity index, function as shown below.

	I
40	index
41	=FaascheQuantityIndex(F41:F43,E41:E43,B41:B43)
42	=FaascheQuantityIndex(D41:D43,C41:C43,B41:B43)
43	=FaascheQuantityIndex(F41:F43,E41:E43,B41:B43)
44	

After entering the Paasche quantity index function, the Excel data sheet should look like the following.

	A	B	C	D	E	F	G	H	I
40	1991	quant91	1992	quant92	1993	quant93	commod	indexyr	index
41	2000	40	2100	50	2500	60	auto	1991	154.032
42	600	30	600	40	700	50	comput	1992	128.571
43	250	10	300	20	300	20	tele	1993	154.032

19.8 FISHER'S IDEAL Quantity Index

Fisher's ideal quantity index is defined as

$$FIQ = \sqrt{(\text{Laspeyres quantity index})(\text{Paasche quantity index})}$$

We can use MINITAB to calculate the FIQ for 1992 and 1993:

```
MTB > let k1=sqrt(129.825*128.571)
MTB > print k1
```

Data Display

K1   129.196

```
MTB > let k2=sqrt(159.649*154.032)
MTB > print k2
```

Data Display

K2   156.815

From the MINITAB output the 1992 FIQ which is contained in k1 is 129.196 and the 1993 FIQ, contained in k2, is 156.815.

19.9 Statistical Summary

In this chapter, we discussed several price and quantity indexes that are used to analyze economic activities.

CHAPTER 20
SAMPLING SURVEYS:
METHODS AND APPLICATIONS

20.1 Introduction

In previous chapters we investigated sampling but only in terms of simple random sampling, in which each potential sample of n members has an equal chance of being chosen. Most of the time this requirement is satisfied by the fact that the sample size is small compared to the population. But when the sample size becomes a large part of the population, some adjustments must be made.

20.2 Random Number Tables

Random tables are very often used to obtain random samples. MINITAB can easily create random tables in response to the *random* command. As shown in Chapters 8 and 9, the *random* command can randomly generate numbers from many different kinds of distributions. The distribution that we will use here to generate random tables is the uniform distribution, because in many applications we are interested in every number having an equal chance of occurring. The uniform distribution has this property.

Suppose we are interested in creating a 10 by 10 table that has numbers randomly generated from 0 to 100. The creation of this table is illustrated below.

```
MTB > random 10 c1-c10;
SUBC> integer 0 100.
MTB > print c1-c10
```

Data Display

Row	C1	C2	C3	C4	C5	C6	C7	C8	C9	C10
1	44	59	3	51	72	78	85	95	72	95
2	8	20	89	40	73	26	49	28	30	15
3	18	90	30	14	62	48	45	0	39	19
4	46	0	28	2	45	73	73	50	48	3
5	44	73	36	93	30	46	31	6	89	46
6	6	81	73	6	64	53	12	13	94	24
7	32	41	96	66	13	46	77	63	13	91
8	32	5	79	45	89	8	90	96	73	15
9	48	11	26	66	75	5	90	73	69	63
10	41	8	48	45	37	55	45	83	84	83

We used the *integer* subcommand instead of the *uniform* subcommand because the *integer* subcommand is a uniform distribution that has only integer numbers. Now let us create another random table, using exactly the same commands, to illustrate that random tables are in fact random.

```
MTB > random 10 c1-c10;
SUBC> integer 0 100.
MTB > print c1-c10
```

Data Display

Row	C1	C2	C3	C4	C5	C6	C7	C8	C9	C10
1	31	39	30	28	7	10	43	0	57	97
2	81	6	68	97	67	18	77	26	60	56
3	91	69	97	46	69	50	20	66	88	53
4	92	53	18	84	93	98	12	90	79	50
5	93	55	89	1	48	84	2	71	45	88
6	1	76	22	53	48	78	81	2	29	74
7	85	49	41	46	3	60	81	22	59	43
8	48	80	95	71	66	5	41	24	15	67
9	32	18	24	79	25	15	18	0	84	76
10	52	11	14	67	4	93	79	47	50	99

Below is the EXCEL code to create a 10 by 10 table that has numbers randomly generated from 0 to 100.

```
'/***********************************************************
'/Purpose:   To generate 100 numbers from 0 to 100 on the
'/           active sheet
'/***********************************************************
Sub Random100()
    Dim rRandom As Range
    Dim rCell As Range

    Set rRandom = ActiveSheet.Range("a1:j10")

    For Each rCell In rRandom.Cells
        Randomize   ' Initialize random-number generator.
        rCell.Value = Rnd * 100
    Next

    rRandom.NumberFormat = "0"
End Sub
```

Below shows a 10 by 10 random table generated by the above EXCEL code.

	A	B	C	D	E	F	G	H	I	J
1	85	29	97	18	29	22	44	77	79	20
2	23	41	87	50	7	37	99	12	49	64
3	44	5	96	97	81	91	62	48	62	21
4	72	44	75	82	87	71	19	75	34	4
5	96	74	13	95	77	3	23	90	90	65
6	39	91	34	58	86	50	84	56	64	14
7	91	86	75	10	4	48	90	37	49	41
8	64	70	98	39	42	60	79	69	26	56
9	92	31	68	57	37	71	88	89	86	95
10	67	53	67	25	50	49	80	87	92	76

20.3 Confidence Interval for the Population Mean

In previous chapters we assumed either that the population size was sufficiently large or the sample was small enough compared to the population for the sample to be random. When this is not the case, we have to make adjustments. One such adjustment is the variance of the sample mean. The adjusted variance of the sample mean is defined as

$$\hat{\sigma}_{\bar{x}}^2 = \frac{s^2}{n} * \frac{N-n}{N-1}$$

As a rule of thumb, if $n/N > 0.05$, we should use the above adjusted sample variance. If $n/N < 0.05$, we should use the calculated sample variance, s^2.

Example 20.3A

Suppose an investment adviser is trying to decide whether a small retirement community consisting of 1,000 residents represents a promising source of potential clients. To determine the potential business, the investment adviser decides to analyze the size of the residents' investment portfolios. A random sample of 75 residents, who were able to respond anonymously, produces a sample mean of $375,000 with a sample standard deviation of $120,000. Calculate the 95% confidence interval for the mean value of the investment portfolio.

The ratio of the sample size to the population size is $n/N = 75/1000 = 0.08$. Since the ratio of the sample size to the population size is greater than 0.05, we should use the adjusted sample variance.

When we do not have to make an adjustment to the variance of the sample mean, the formula for the confidence interval would be

$$\bar{x} - z_{\alpha/2}(\sigma/\sqrt{n}) < \mu < \bar{x} + z_{\alpha/2}(\sigma/\sqrt{n})$$

With the adjustment to the variance of the sample mean, the confidence interval formula is

$$\bar{x} - z_{\alpha/2}\hat{\sigma}_{\bar{x}} < \mu < \bar{x} + z_{\alpha/2}\hat{\sigma}_{\bar{x}}$$

We will create a macro to calculate the adjusted variance of the sample mean and the confidence interval for a population mean that came from a large sample. To do this, we will borrow a lot from the *zint* macro in Chapter 10 to program the macro *adjzint*. The analysis of the *adjzint* macro is very similar to that of the *zint* macro.

The use of the macro *adjzint* to solve Example 20.3A is shown below.

328   CHAPTER 20

```
MTB > %adjzint
Executing from file: adjzint.MAC
this macro calculates the adjusted sample variance
and then finds a confidence interval for a sample mean
based on the adjusted sample variance

What is the population size?
DATA> 1000
What is the sample size?
DATA> 75
What is the sample standard deviation?
DATA> 120000
The adjusted sample variance is

Data Display

C60
    177777778

The adjusted standard deviation is
```

Data Display

```
C61
    13333.3

What is the mean?
DATA> 375000
What is the desired percentage confidence interval?
DATA> .95
The confidence interval is:
```

Data Display

```
Row    lowint    highint

 1     348867    401133
```

From the *adjzint* macro we can see that the 95% confidence interval is from $348,867 to $401,133. The set of commands in the *adjzint* macro is shown below.

Adjzint Macro

```
gmacro
noecho
erase c50-c80
note this macro calculates the adjusted sample variance
note and then finds a confidence interval for a sample mean
note based on the adjusted sample variance
note
note What is the population size?
set 'terminal' c50;
nobs=1.
end
note What is the sample size?
set 'terminal' c51;
nobs=1.
end
note What is the sample standard deviation?
set 'terminal' c52;
nobs=1.
end
#calculate variancee
let c60=((c52**2)/c51)* ((c50-c51)/(c50-1))
let c61=sqrt(c60)
note The adjusted sample variance is
print c60
note The adjusted standard deviation is
print c61
note
note What is the mean?
set 'terminal' c50;
nobs=1.
end
note What is the desired percentage confidence interval?
set 'terminal' c53;
nobs=1.
end
let k50=c50
let k53=c53
#k53 has the confidence interval percentage
let k55=(1+k53)/2
invcdf k55 k56;
normal 0 1.
#calculate intervals
let c71=k50+k56*c61
let c70=k50-k56*c61
name c70 'lowint' c71 'highint'
note The confidence interval is:
print c70 c71
erase c50-c80
endmacro
```

20.4 Confidence Interval for the Population Proportion

We will also need to adjust the variance of the sample proportion when the sample relative to the population is not so small. The adjusted variance for a sample proportion is

$$\hat{\sigma}_{\bar{x}}^2 = \frac{\hat{p}(1-\hat{p})}{n} * \frac{N-n}{N-1}$$

Example 20.4A

Suppose we want to determine the proportion of college-bound high school seniors in a class of 500. A survey of 30 randomly selected students reveals that 19 will be attending college. Create a 90% confidence interval for the population proportion *p*.

The ratio of the sample size to the population size is 30/500 = 0.06. Since this is greater than .05, we will have to make an adjustment to the sample variance. To do this, we will create a macro called *adjpint* to calculate both the adjusted variance and the confidence interval for a population proportion. Because we will borrow a lot from the *pint* macro in Chapter 10 to program the macro *adjpint*, the analysis of the *adjpint* macro is very similar to that of the *pint* macro.

```
MTB > %adjpint
Executing from file: adjpint.MAC
this macro calculates the adjusted sample variance
and then finds a confidence interval for a sample proportion
based on the adjusted sample variance

What is the population size?
DATA> 500
What is the sample size?
DATA> 30
What is the proportion?
DATA> .6333
The adjusted sample variance is
```

Data Display
```
C60
    0.0072912
```

The adjusted standard deviation is

Data Display
```
C61
    0.0853883
```

```
What is the desired percentage confidence interval?
DATA> .90
The confidence interval is:
```

Data Display

```
Row     lowint      highint

 1     0.492849    0.773751
```

From the *adjpint* macro we see that the 90% confidence interval is from 49.2849% to 77.3751%. Following are the commands in the *adjpint* macro.

Adjpint Macro

```
Gmacro
noecho
erase c50-c80
# this macro is used for chapter 20
note this macro calculates the adjusted sample variance
note and then finds a confidence interval for a sample proportion
note based on the adjusted sample variance
note
note What is the population size?
set 'terminal' c50;
nobs=1.
end
note What is the sample size?
set 'terminal' c51;
nobs=1.
end
note What is the proportion?
set 'terminal' c52;
nobs=1.
end
let c60=((c52*(1-c52))/c51) * ((c50-c51)/(c50-1))
let c61=sqrt(c60)
note The adjusted sample variance is
print c60
note The adjusted standard deviation is
print c61
note
note What is the desired percentage confidence interval?
set 'terminal' c53;
nobs=1.
end
let k53=c53
```

```
#k53 has the confidence interval percentage
let k55=(1+k53)/2
invcdf k55 k56;
normal 0 1.
let c71=c52+k56*c61
let c70=c52-k56*c61
name c70 'lowint' c71 'highint'
note The confidence interval is:
print c70 c71
erase c50-c80
endmacro
```

20.5 Determining Sample Size

Up to now we have obtained the sample size first and then found the standard deviation. What if we did the reverse and knew a specific standard deviation but did not know the sample size? The formula to find the required sample size for a required standard deviation would be

$$n = \frac{N\sigma^2}{(N-1)\hat{\sigma}_{\bar{x}}^2 + \sigma^2}$$

Example 20.5A

Crow Company's accountant decides that the best way to determine the company's mean accounts receivable is to take a simple random sample of the 1,025 accounts. Assume that the population variance, σ^2, is $2,425. What sample size should the accountant take if she would like to have a level of precision of $75?

We will create a macro called *sampsize* to calculate the sample size for this required precision.

The *sampsize* macro is shown below.

Sampsize Macro

```
gmacro
noecho
erase c50-c80
# this macro is used for chapter 20
note This macro calculates the required sample size
note for a specific standard deviation.
note What is the population size?
set 'terminal' c50;
nobs=1.
end
note What is the sample variance?
set 'terminal' c52;
nobs=1.
end
note What is the population variance?
set 'terminal' c53;
nobs=1.
end
let c60=(c50*c53)/((c50-1)*c52+c53)
note
note The required sample size is
print c60
erase c50-c80
endmacro
```

Now let's see how we would use *sampsize* in the Crow Company case.

```
MTB > %sampsize
Executing from file: sampsize.MAC
This macro calculates the required sample size
for a specific standard deviation.
What is the population size?
DATA> 1025
What is the sample variance?
DATA> 75
What is the population variance?
DATA> 2425

The required sample size is
```

Data Display

```
C60
    31.3743
```

From the *sampsize* macro we can see that a sample of 32 accounts will produce the desired result.

Example 20.5B

Use the information in Example 20.5A, but assume the accountant would like a sample variance of $50.

```
MTB > execute 'sampsize'
This macro calculates the required sample size
for a specific standard deviation.
What is the population size?
DATA> 2425
What is the sample variance?
DATA> 50
What is the population variance?
DATA> 2425

The required sample size is

C60
  46.352
```

From the *sampsize* macro output we can see that 47 is the required sample size for a sample variance of $50. Notice that as the sample variance decreases, the sample size increases.

20.6 Statistical Summary

In this chapter we first discussed how random number tables can be generated. Then we discussed samples that have a ratio of sample size to population size greater than 0.05. This is important because in such a case we have to make some adjustments to make the estimate unbiased.

We also learned how to determine the required sample size for a desired sample variance.

CHAPTER 21
STATISTICAL DECISION THEORY:
METHODS AND APPLICATIONS

21.1 Introduction

In Chapters 1 to 20 we have been looking at classical statistics, in which we focused on estimation, constructing intervals, and hypothesis testing. In this chapter we will briefly look at another branch of statistics called *statistical decision theory*. In statistical decision theory we look at feasible alternative courses of action in an uncertain environment.

In decision making there are four elements: actions, states of nature, outcomes, and probabilities.

1. The choices available to the decision maker are called *actions*.
2. The uncertain elements in a problem are referred to as *states of nature*.
3. An *outcome* is a consequence of each combination of an action and event (state of nature).
4. *Probability* is the chance that an event will occur.

21.2 Decisions Based On Extreme Values

Most of the decision rules discussed in this chapter require the specification of probabilities for the state of nature. However, the *maximum criterion* does not require the use of probabilities. The maximum criterion first considers the worst possible outcome for each action and then chooses the highest of the minimum payoffs.

Example 21.2A

Suppose a firm can produce any one of four products, 1, 2, 3, and 4. The four products, then, are the possible actions. The state of nature (which in this case is the state of the economy) has three possibilities: recession, flat, and boom. The payoff, or outcome, is whatever profits result. The information is presented in the table below.

	States of Nature		
Product	Recession	Flat	Boom
1	–$10	–$5	$20
2	–$15	$0	$15
3	–$7	–$1	$25
4	–$3	–$2	$17

Use MINITAB to find out which of the products we should produce using the maximum criterion. The worksheet should look like the following after entering the data and naming the columns.

	C1	C2	C3	C4	C5	C6
	product	recess	flat	boom	mimpay	
1	1	-10	-5	20	-10	
2	2	-15	0	15	-15	
3	3	-7	-1	25	-7	
4	4	-3	2	17	-3	
5						
6						
7						
8						

The '*mimpay*' column in the worksheet is the maximum loss of each product.

```
MTB > rmin c2-c4 c5
MTB > print c1-c5
```

Data Display

```
Row   product   recess   flat   boom   mimpay

  1       1      -10      -5     20     -10
  2       2      -15       0     15     -15
  3       3       -7      -1     25      -7
  4       4       -3       2     17      -3
```

```
MTB > maximum 'mimpay'
```

Column Maximum

```
Maximum of mimpay = -3.0000
```

21.3 Expected Monetary Value

The monetary value of a particular state of nature is calculated by multiplying the probability of the action by the payoff of that particular action. The best alternative is the one that maximizes the expected monetary value (EMV).

Example 21.3A

For example, suppose economists estimate the following probabilities about the economy and certain products.

	States of Nature		
	Recession	Flat	Boom
Probability of Occurring	0.2	0.5	0.3
Product 1	–$10	–$5	$20
Product 2	–$15	$0	$15
Product 3	–$7	–$1	$25
Product 4	–$3	–$2	$17

The EMV for each product would be:

$$EMV_1 = (0.2)(-20) + (0.5)(0) + (0.3)(15) = \$0.5$$
$$EMV_2 = (0.2)(-5) + (0.5)(-2) + (0.3)(5) = -\$0.5$$
$$EMV_3 = (0.2)(-10) + (0.5)(1) + (0.3)(0.25) = \$6.0$$
$$EMV_4 = (0.2)(-1) + (0.5)(5) + (0.3)(0) = \$2.3$$

To do the calculation in MINITAB, the worksheet should look like the following before entering any commands.

	C1	C2	C3	C4	C5	C6
	prod1	prod2	prod3	prod4	probab	emv
1	-20	-5	-10	-1	0.2	
2	0	-2	1	5	0.5	
3	15	5	25	0	0.3	
4						

```
MTB > let 'emv'(1) = sum('prod1'*'probab')
MTB > let 'emv'(2) = sum('prod2'*'probab')
MTB > let 'emv'(3) = sum('prod3'*'probab')
MTB > let 'emv'(4) = sum('prod4'*'probab')
MTB > print c1-c6
```

Data Display

```
Row   prod1   prod2   prod3   prod4   probab   emv

 1     -20     -5      -10     -1      0.2      0.5
 2       0     -2        1      5      0.5     -0.5
 3      15      5       25      0      0.3      6.0
 4                                              2.3
```

```
MTB > maximum 'emv'
```

Column Maximum

```
Maximum of emv = 6.0000
MAXIMUM =        6.0000
```

From the MINITAB output we can see that the highest EMV is 6, which belongs to product 3.

21.4 Bayes Strategies

We will now look at Bayesian strategies, which consider the probability for each state of nature. These strategies use the Bayes theorem, which we will derive.

The formula for the one-event Bayes theorem is

$$P(Y|X) = \frac{P(Y \cap X)}{P(X)} \tag{21.1}$$

We can read the left side of the equation as, 'What is the probability of event Y given event X?' The right side of the equation will be read as the probability of the intersection of event Y and event X divided by the probability of event X. Graphically, it would look like the following

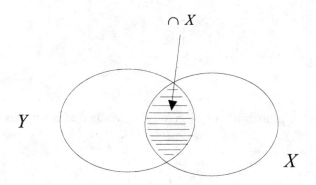

From the above graph we can also state

$$P(X|Y) = \frac{P(Y \cap X)}{P(Y)} \qquad (21.2)$$

If we multiply both sides of Equation (21.2) by $P(Y)$, we would get

$$P(Y \cap X) = P(X|Y)P(Y) \qquad (21.3)$$

If we multiply both sides of Equation (21.1) by $P(X)$, we would get

$$P(Y \cap X) = P(X|Y)P(X) \qquad (21.4)$$

From Equations (21.3) and (21.4) we would get

$$P(X|Y)P(Y) = P(X|Y)P(X) \qquad (21.5)$$

Two events would look like what is seen in the following.

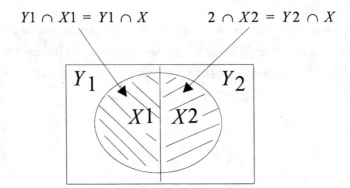

We can ask, 'What is the probability of $y1$ given x?'. Mathematically, this would be expressed as follows.

$$P(Y1|X) = \frac{P(Y1 \cap X)}{P(Y1 \cap X) + P(Y2 \cap X)} \qquad (21.6)$$

We can also ask, 'What is the probability of $y2$ given x?' Mathematically, this would be

$$P(Y2|X) = \frac{P(Y2 \cap X)}{P(Y1 \cap X) + P(Y2 \cap X)} \qquad (21.7)$$

Three events would look like:

We can ask, 'What's the probability of $y1$ given x?'. Mathematically, the question would be

$$P(Y1|X) = \frac{P(Y1 \cap X)}{P(Y1 \cap X) + P(Y2 \cap X) + P(Y3 \cap X)} \qquad (21.8)$$

We can ask, 'What's the probability of $y2$ given x?'. Mathematically, it would be

$$P(Y2|X) = \frac{P(Y2 \cap X)}{P(Y1 \cap X) + P(Y2 \cap X) + P(Y3 \cap X)} \qquad (21.9)$$

We can also ask, 'What's the probability of $y3$ given x?'. Mathematically, it would be

$$P(Y3|X) = \frac{P(Y3 \cap X)}{P(Y1 \cap X) + P(Y2 \cap X) + P(Y3 \cap X)} \qquad (21.10)$$

Following the same logic as for one, two, and three events, we can write n events as

$$P(Yi|X) = \frac{P(Yi \cap X)}{P(Y1 \cap X) + P(Y2 \cap X) + ... + P(Yn \cap X)} \qquad (21.11)$$

Equation (21.11) is the generalized Bayes theorem. But the more popular form is to substitute equation (21.3) and equation (21.4) into (21.11) to get

$$P(Yi|X) = \frac{P(X|Yi)P(Yi)}{P(X|Y1)P(Y1) + P(X|Y2)P(Y2) + ... + P(X|Yn)P(Yn)} \qquad (21.12)$$

Some books like to reduce equation (21.12) to the following form

$$P(Yi|X) = \frac{P(X|Yi)P(Yi)}{\sum_{j=1}^{m} P(X|Yj)P(Yj)} \qquad (21.13)$$

Example 21.4A

Suppose that macroeconomists are hired to predict interest rates. Past results for economic prognosticators are presented in the following table.

342   CHAPTER 21

Belief	Up	Down
Strong Credit Market	0.6	0.3
Weak Credit Market	0.4	0.7

It is believed that the probability that rates will rise is 0.7 and the probability of lower rates is 0.3.

What is the probability that interest rates will rise, given a strong credit market assessment by the economists?

To answer this question we can use equation (21.10), the two event Bayesian form. The answer would be

$$P(\text{up}|\text{strong market}) = \frac{P(\text{strong market}|\text{up})P(\text{up})}{P(\text{strong market}|\text{up})P(\text{up}) + P(\text{strong market}|\text{down})P(\text{down})}$$

$$= \frac{0.60(0.70)}{0.6(0.7) + 0.3(0.3)} = \frac{0.42}{0.51} = 0.82$$

MINITAB does not have a Bayes' theorem command; therefore, we will create our own macro called *bayesian* to calculate the Bayes' theorem.

Bayesian Macro

```
gmacro
noecho
# This macro is used in chapter 21

note This macro calculates the bayes theorem
note
note How many events in the bayes calculation?
read 'terminal' c50;
nobs=1.
end
Let k50=c50
Note What is the conditional probability, and probability of events?
read 'terminal' c60 c61;
nobs = k50.
end
let k60=sum(c60*c61)
let c65=c60*c61/k60
note The bayesian probability for each event is
name c65 'bay prob' c60 'con prob' c61 'prb evt'
print c60 c61 c65
endmacro
```

```
MTB > %bayesian
Executing from file: bayesian.MAC
This macro calculates the bayes theorem

How many events in the bayes calculation?
DATA> 2
     1 rows read.
What is the conditional probability, and probability of events?
DATA> .6 .7
DATA> .3 .3
     2 rows read.
The bayesian probability for each event is
```

Data Display

Row	con prob	prb evt	bay prob
1	0.6	0.7	0.823529
2	0.3	0.3	0.176471

From our *bayesian* output we can see that the probability of the interest rate going up in a strong market is 82.3529%.

21.5 Statistical Summary

In this chapter we briefly looked at a branch of statistics called *statistical decision theory*. Some of the statistical decision theory methods we examined were the maximum criterion method, expected monetary value, and Bayes strategies.

CHAPTER 22
USING MICROSOFT EXCEL TO
ESTIMATE ALTERNATIVE OPTION PRICING MODELS[1]

21.1 Introduction

This chapter shows how Microsoft Excel can be used to estimate call and put options for (a) Black-Scholes model for individual stock, (b) Black-Scholes model for stock indices, and (c) Black-Scholes model for currencies.

22.2 Option Model for Individual Stock

The call option formula for the individual stock can be defined as

$$C = SN(d_1) - Xe^{-r(T)}N(d_2) \qquad (22.1)$$

where

$$d_1 = \frac{\ln\left(\frac{S}{X}\right) + \left(r + \frac{1}{2}\sigma^2\right)T}{\sigma\sqrt{T}}$$

$$d_2 = \frac{\ln\left(\frac{S}{X}\right) + \left(r - \frac{1}{2}\sigma^2\right)T}{\sigma\sqrt{T}} = d_1 - \sigma\sqrt{T}$$

C = price of the call option
S = current price of the stock
X = exercise price of the option
e = 2.71828...
r = short-term interest rate (T-Bill rate) = R_f
T = time to expiration of the option, in years
$N(d_i)$ = value of the cumulative standard normal distribution (i = 1,2)
σ^2 = variance of the stock rate of return

---

[1] This chapter was written by Professor Cheng F. Lee and Dr. Ta-Peng Wu of Rutgers University.

The put option formula for the formula can be defined as

$$P = Xe^{-r(T)}N(-d_2) - SN(-d_1) \qquad (22.2)$$

where
P = price of the put option.

The other notations have been defined in equation (22.1)

Following example 7C.1 of Lee et. al. (2000, pp. 290-91), assume that S = 42, X = 40, r = 0.1, σ = 0.2 and T = 0.5. The following shows how to set up Microsoft Excel to solve the problem.

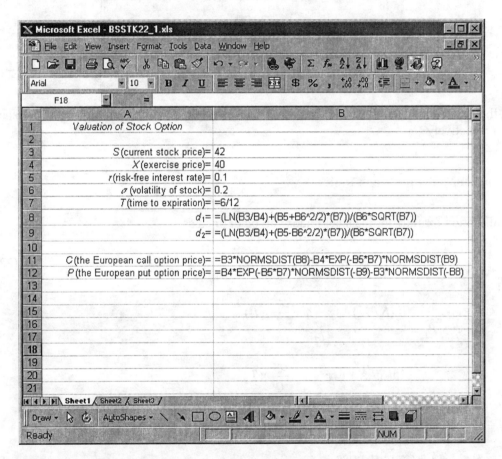

The following shows the answer to the problem in Microsoft Excel.

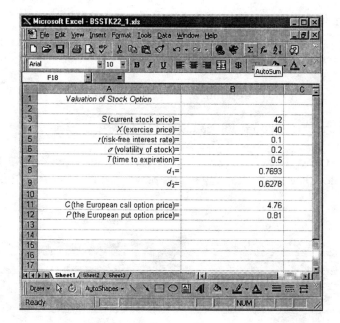

From the Excel output, we find that the price of the call option and the put option is $4.76 and $0.81 respectively

22.3 Option Model for Stock Indices

The call option formula for the stock index can be defined as

$$C = Se^{-q(T)}N(d_1) - Xe^{-r(T)}N(d_2) \tag{22.3}$$

where

$$d_1 = \frac{\ln(S/X) + (r - q + \frac{\sigma^2}{2})(T)}{\sigma\sqrt{T}}$$

$$d_2 = \frac{\ln(S/X) + (r - q - \frac{\sigma^2}{2})(T)}{\sigma\sqrt{T}} = d_1 - \sigma\sqrt{T}$$

q = dividend yield
S = value of index
X = exercise price
r = short-term interest rate (T-Bill rate) = R_f
T = time to expiration of the option, in years
$N(d_i)$ = value of the cumulative standard normal distribution (i = 1,2)
σ^2 = variance of the stock rate of return

The put option formula for the stock index can be defined as

$$P = Xe^{-r(T)}N(-d_2) - Se^{-q(T)}N(-d_1) \qquad (22.4)$$

where

P = the price of the put option.

The other notations have been defined in equation (22.3).

Following example 19A.1 of Lee et. al. (2000, pp. 875-76), assume that S = 950, X = 900, r = 0.06, σ = 0.15, q = 0.03, and T = 2/12. The following shows how to set up Microsoft Excel to solve the problem.

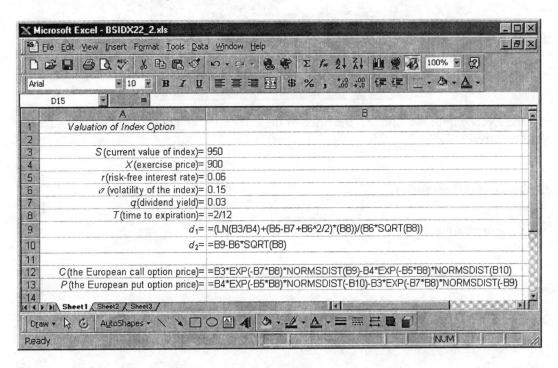

The following shows the answer to the problem in Microsoft Excel.

348   Chapter 22

	A	B
1	Valuation of Index Option	
3	S(current value of index)=	950.00
4	X(exercise price)=	900.00
5	r(risk-free interest rate)=	0.06
6	σ (volatility of the index)=	0.15
7	q(dividend yield)=	0.03
8	T(time to expiration)=	0.17
9	d_1=	0.9952
10	d_2=	0.9339
12	C(the European call option price)=	59.22
13	P(the European put option price)=	5.01

From the Excel output, we find that the price of the call option and the put option is 59.22 and 5.01 respectively.

22.4 Option for Currencies

The call option formula for the currency can be defined as

$$C = Se^{-r_f(T)}N(d_1) - Xe^{-r(T)}N(d_2)$$

where

$$d_1 = \frac{\ln(S/X) + (r - r_f + \frac{\sigma^2}{2})(T)}{\sigma\sqrt{T}}$$

$$d_2 = \frac{\ln(S/X) + (r - r_f - \frac{\sigma^2}{2})(T)}{\sigma\sqrt{T}} = d_1 - \sigma\sqrt{T}$$

S = spot exchange rate,
r = short-term interest rate (T-Bill rate) = R_f
r_f = foreign risk free rate
X = exercise price
T = time to expiration of the option, in years
$N(d_i)$ = value of the cumulative standard normal distribution (i = 1,2)
σ = standard deviation of spot rate

The put option formula for the currency can be defined as

$$P = Xe^{-r(T)}N(-d_2) - Se^{-r_f(T)}N(-d_1) \tag{22.6}$$

where

P = the price of the put option.

The other notations have been defined in equation (22.5).

Following example 19A.2 of Lee et. al. (2000, pp. 876-77), assume that S = 130, X = 125, r = 0.06, r_f = 0.02, σ = 0.15, and T=4/12. The following shows how to set up Microsoft Excel to solve the problem.

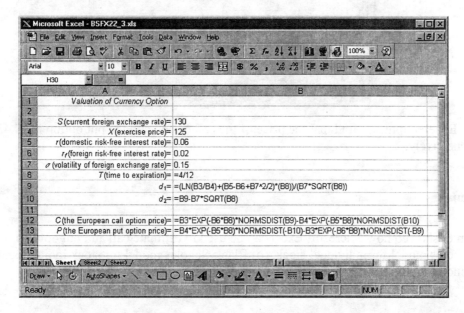

The following shows the answer to the problem in Microsoft Excel.

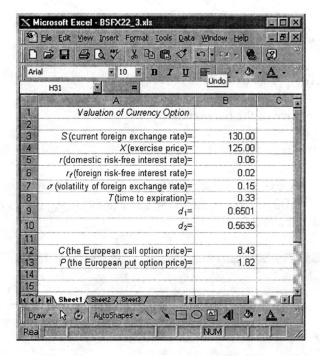

From the Excel output, we find that the price of the call option and the put option is $8.43 and $1.82 respectively.

22.5 Summary

This chapter showed how Microsoft Excel can be used to estimate call and put options for (a) Black-Scholes model for individual stock, (b) Black-Scholes model for stock indices, and (c) Black-Scholes model for currencies.

APPENDIX A
REFERENCES

A.1 Microsoft Excel References

I feel that one of the best book on Microsoft Excel programming is by John Walkenbach.

Microsoft Excel 2000 Power Programming with VBA, John Walkenbach, IDG Books WorldWide, New York, 1999.

The Internet has made access to information very easy. They following are excellent Internet Sites about Microsoft Excel Development.

Microsoft Excel Home Page
http://www.microsoft.com/excel

John Walkenbach's Excel Page
http://www.j-walk.com/ss

Stephen Bullen's Excel Page
http://www.bmsltd.co.uk/excel

Chip Pearson's Excel Page
http://www.cpearson.com/excel

A.2 Statisical References

Statistics for Business and Financial Economics, Cheng-Few Lee, John C. Lee, Alice C. Lee, Second Edition, World Scientific, Singapore and New Jersey, 2000.
Statistics for Managers using Microsoft Excel, David M. Levine, Mark L. Berenson, David Stephan, Prentice Hall, New Jersey, 1998.

A.3 MINITAB References

http://www.minitab.com/resources/ctl/Business.htm

INDEX

#, 153
add-in, 6
adjpint, 331
adjzint, 329
alt-F11,34
analysis tools pack, 189
anova, 190
aovoneway, 187
arithmetic mean, 39
autocorrelation, 262
average, 41
Bayesian, 342
binomdist,79
binomial,121,79
binomial distribution, 121
binomial random variable, 78
bintreg, 224
box-and-wisker plot, 30
boxplot, 30,261
cdf, 72
chisquare,132
center, 49
central limit theorem, 107,136
chi-square distribution, 131
coefficient of skewness, 51
comb, 65
combin, 66
combination,64
confidence,153
continuous random variable, 85
correl, 214
correlation,266
counts, 19, 24
culmuniformpdf, 90
cumpercents, 25
describe, 43,150,172
discrete, 73
discrete random variable, 70
dotplot, 27, 177, 260

drawboxplot, 32
durbin-watson, 262
exponential, 136
exponential distribution, 134
expsmth, 290
F distribution, 133, 139
fisheridealpriceindex, 317
formula bar, 41
heteroscedasticity, 261
histrogram, 28
holtreg, 298
holtzero, 297
hypothesis testing, 168
increment, 28
invcdf, 239
kurskal-wallis, 274
laspeyrespriceindex, 311
laspeyresquantityindex, 320
lognormal, 116, 101
lognormal distribution, 100, 116
ma.mac, 285
mann-whitney, 271
maximum, 277, 285
mean, 39
median, 42
noecho, 153
normal, 113
normal distribution, 91, 113
normalpdf, 96
paaschepriceindex, 314
paaschequantityindex,322
pdf, 79, 82
percents, 25
perm, 59
permut, 63
permutation, 59
phyp2, 179
pint, 161
poisson, 124

352

poisson random variable, 82
pooled, 177
population, 56
predict, 226, 245
priceindex, 307
print, 200
probability, 57
professional graphs, 28
p-value, 171
random100
random, 108, 121, 113, 116, 124, 136, 324
rank, 281
regress, 222, 226, 245, 252
regression, 207, 236
same, 177
sample, 56
sample size, 333
set, 54
skewness, 53
stack, 281
standard deviation, 44
standard graphs, 28
standard normal distribution, 102
start, 28
stdev, 44
stem and leaf, 29
stepwise, 268
stepwise regression, 248
sumproduct, 74
sumxmy2, 263
t distribution, 129, 145
table, 58
tally, 25
template, 154
tint, 159
trend, 288
tukey, 194
twosample, 177
uniform distribution, 85, 109
uniformpdf, 88
var, 46
variance, 45
visual basic editor, 34

wilcoxon, 277
wtest, 278
xint, 166
z score, 48
zint, 151
zinterval, 150
ztest, 169